MY
REVISION
NOTES

T-LEVELS
THE NEXT LEVEL QUALIFICATION

ONSITE CONSTRUCTION

Mike Jones
Stephen Jones
Tom Leahy

'T-LEVELS' is a registered trade mark of the Department for Education. 'T Level' is a registered trade mark of the Institute for Apprenticeships and Technical Education. The T Level Technical Qualification is a qualification approved and managed by the Institute for Apprenticeships and Technical Education.

The Publishers would like to thank the following for permission to reproduce copyright material.

Photo credits
pp.20–1 *Table* 1.4 *t–b* © NESRUDHEEN/stock.adobe.com, © Daseaford/stock.adobe.com, © Viktorijareut/stock.adobe.com, © Viktorijareut/stock.adobe.com, © Ricochet64/stock.adobe.com; **pp.21–2** *Table* 1.5 *all* © Ricochet64/stock.adobe.com, *except Corrosive safety sign* © Olando/stock.adobe.com; **p.96** © Jason/stock.adobe.com

Although every effort has been made to ensure that website addresses are correct at time of going to press, Hodder Education cannot be held responsible for the content of any website mentioned in this book. It is sometimes possible to find a relocated web page by typing in the address of the home page for a website in the URL window of your browser.

Hachette UK's policy is to use papers that are natural, renewable and recyclable products and made from wood grown in well-managed forests and other controlled sources. The logging and manufacturing processes are expected to conform to the environmental regulations of the country of origin.

Orders: please contact Hachette UK Distribution, Hely Hutchinson Centre, Milton Road, Didcot, Oxfordshire, OX11 7HH. Email education@hachette.co.uk Telephone: +44 (0)1235 827827. Lines are open from 9 a.m. to 5 p.m., Monday to Friday. You can also order through our website: www.hoddereducation.co.uk

ISBN: 978 1 3983 8452 1

Cover photo © Monkey Business - stock.adobe.com

Illustrations by Integra Software Services Pvt. Ltd.

Typeset in Integra Software Services Pvt. Ltd., Pondicherry, India

Printed in Spain

A catalogue record for this title is available from the British Library.

Get the most from this book

Everyone has to decide their own revision strategy, but it is essential to review your work, learn it and test your understanding. These Revision Notes will help you do that in a planned way, topic by topic. Use this book as the cornerstone of your revision and don't hesitate to write in it – personalise your notes and check your progress by ticking off each section as you revise.

Tick to track your progress

Use the revision planner on pages 4–6 to plan your revision, topic by topic. Tick each box when you have:
+ revised and understood a topic
+ tested yourself
+ practised the exam questions and checked your answers online.

You can also keep track of your revision by ticking off each topic heading in the book. You may find it helpful to add your own notes as you work through each topic.

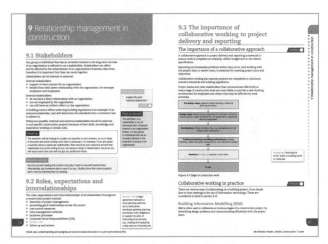

Features to help you succeed

Exam tips

Expert tips from the authors are given throughout the book to help you polish your exam technique and maximise your chances in the exam.

Typical mistakes

The authors identify the typical mistakes students make and explain how you can avoid them.

Now test yourself

These short, knowledge-based questions provide the first step in testing your learning. Answers are available online at www.hoddereducation.co.uk/myrevisionnotesdownloads

Key terms

Clear, concise definitions of key terms are provided where they first appear.

Revision activities

These activities will help you revise each topic in an interactive way.

Exam-style questions

Practice exam-style questions are provided for each topic. Use them to consolidate your revision and practise your exam skills. The answers are available online at www.hoddereducation.co.uk/myrevisionnotesdownloads

My revision planner

Check your understanding and progress at **www.hoddereducation.co.uk/myrevisionnotes**

My revision planner

REVISED   TESTED   EXAM READY

9 Relationship management in construction

10 Digital technology in construction

11 Construction commercial/business principles

My revision planner

Check your understanding and progress at **www.hoddereducation.co.uk/myrevisionnotes**

Countdown to my exams

From September

+ Attend class in person or via the internet if necessary.
+ Listen and enjoy the subject; make notes.
+ Make friends in class and discuss the topics with them.
+ Watch the news.

6–8 weeks to go

+ Start by looking at the specification – make sure you know exactly what material you need to revise and the style of the exam. Use the revision planner on pages 4–6 to familiarise yourself with the topics.
+ Organise your notes, making sure you have covered everything on the specification. The revision planner will help you group your notes into topics.
+ Work out a realistic revision plan that will allow you time for relaxation. Set aside days and times for all the subjects that you need to study and stick to your timetable.
+ Set yourself sensible targets. Break your revision down into focused sessions of around 40 minutes, divided by breaks. These Revision Notes organise the basic facts into short, memorable sections to make revising easier.

REVISED ◯

2–6 weeks to go

+ Read through the relevant sections of this book and refer to the exam tips, exam checklists, typical mistakes and key terms. Tick off the topics as you feel confident about them. Highlight those topics you find difficult and look at them again in detail.
+ Test your understanding of each topic by working through the 'Now test yourself' questions in this book. Look up the answers online at **www.hoddereducation.co.uk/ myrevisionnotesdownloads**
+ Make a note of any problem areas as you revise, and ask your teacher to go over these in class.
+ Look at past papers. They are one of the best ways to revise and practise your exam skills. Write or prepare planned answers to the exam-style questions provided in this book. Check your answers online at **www.hoddereducation.co.uk/ myrevisionnotesdownloads**
+ Use the revision activities to try out different revision methods. For example, you can make notes using mind maps, spider diagrams or flash cards.
+ Track your progress using the revision planner and give yourself a reward when you have achieved your target.

REVISED ◯

One week to go

+ Try to fit in at least one more timed practice of an entire past paper and seek feedback from your teacher, comparing your work closely with the mark scheme.
+ Check the revision planner to make sure you haven't missed out any topics. Brush up on any areas of difficulty by talking them over with a friend or getting help from your teacher.
+ Attend any revision classes put on by your teacher. Remember, your teacher is an expert at preparing people for exams.

REVISED ◯

The day before the exam

+ Flick through these Revision Notes for useful reminders, for example the exam tips, exam checklists, typical mistakes and key terms.
+ Check the time (is it morning or afternoon?) and place of your exam. Keep in touch with other students in your class.
+ Make sure you have everything you need for the exam – pens, highlighters and water.
+ Allow some time to relax and have an early night to ensure you are fresh and alert.

REVISED ◯

My exams

Paper 1

Date:...

Time: ....

Location: ....

Paper 2

Date:...

Time: ....

Location: ....

My Revision Notes: Onsite Construction T Level

1 Health and safety in construction

1.1 Construction legislation and regulations

The role of legislation and regulations in the construction industry

REVISED

Health and safety legislation and regulations are laws established to protect the health, safety and welfare of people who may be affected by work activities.

The Health and Safety Executive (HSE) is an independent regulator that enforces health and safety legislation and regulations in the UK.

> **Legislation** Current primary laws, sometimes known as Acts, created by UK legislative bodies (the UK Parliament, Scottish Parliament, Welsh Parliament and Northern Ireland Assembly)
>
> **Regulations** Secondary laws made under the authority of the UK legislative bodies that created the primary laws; formal guidelines used to apply the principles of primary laws

Exam tip

It is unlikely that an examiner will expect you to know the dates when health and safety legislation and regulations were last revised. However, you should know the abbreviations for key Acts and regulations, for example the Health and Safety at Work etc. Act 1974 (HASAWA).

How legislation impacts employers, employees and construction projects

REVISED

The primary health and safety legislation in the UK is the Health and Safety at Work etc. Act 1974 (HASAWA). All employers and employees have responsibilities under the HASAWA to protect people from any harm that may be caused by work activities. The main objectives of HASAWA are to:

+ secure the health, safety and welfare of people at work
+ protect people other than those at work from risks to health and safety arising out of or in connection with work activities
+ control the possession and use of highly flammable, explosive and dangerous substances.

Regulations relating to the provision of welfare facilities during construction work

REVISED

Under HASAWA, employers have a duty to provide welfare facilities for employees at their place of work. The Construction (Design and Management) (CDM) Regulations 2015 outline the minimum facilities that should be provided:

+ drinking water
+ toilets
+ washing facilities
+ rest facilities with heating
+ changing rooms with lockers, seating and facilities to dry and store clothing (separate rooms must be provided for men and women)
+ facilities for pregnant women or nursing mothers to rest lying down.

Exam tip

You will be expected to describe the difference between health and safety legislation (for example HASAWA) and regulations (for example RIDDOR).

Bodies responsible for maintaining and updating legislation and regulations

Health and safety legislation is regularly reviewed and updated to reflect changes in the construction industry. People with legal duties must ensure that the information they are using is current and correct. These changes are often made with guidance and support from:
+ employers
+ unions
+ trade associations
+ professional bodies.

The most up-to-date guidance about health and safety legislation and regulations can be found at the HSE's website www.hse.gov.uk/guidance/index.htm.

Exam tip

The examiner will expect you to be able to list some trade associations related to onsite construction.

The implications of not adhering to legislation and regulations

When legislation and regulations are not adhered to by duty holders in the workplace, there is an increased risk to workers and others of a near-miss incident, an injury, ill health or death.

When people suffer loss or injury because of an accident at work, they may seek compensation.

Failure to comply with statutory law is a criminal offence.

HSE inspectors have the power to enforce health and safety law by:
+ entering a workplace without notice
+ investigating when a complaint has been made or an accident has occurred
+ speaking to employers and workers
+ examining equipment and machinery
+ taking samples, for example of sound and dust levels
+ taking photographs and measurements
+ making copies of records or other documentation
+ removing substances and dismantling and removing articles.

If an HSE inspector believes that an employer has breached the law, they may issue:
+ a simple caution
+ an improvement notice
+ a prohibition notice.

Failure to comply with improvement or prohibition notices can result in prosecution, fines and imprisonment.

Duty holders People with legal responsibilities under health and safety law

Statutory law Written law made by the UK Parliament; also known as an Act of Parliament

Improvement notice Legal document issued by the HSE to an employer, instructing them to put right within a specific period of time any health and safety faults identified

Prohibition notice Legal document issued by the HSE to an employer that prevents work from continuing when there has been a serious breach of the law and people are at risk of immediate harm

Statutory and non-statutory documents in construction

Legislation comprises Acts of Parliament and regulations (statutory legislation) which have legal status and must be complied with.

However, there are also many non-statutory guidance documents that offer advice on good practice and compliance with the law, but unless stated they do not need to be followed.

One example of this is an Approved Code of Practice (ACOP).

Approved Code of Practice (ACOP) Document providing advice and guidance on how to comply with health and safety law, published by the HSE

Regulations and guidance documents

The overarching guidance documents (ACOPs) for working in the construction sector are covered in section 1.3.

The main regulations that control health, safety and welfare in construction are:

+ Control of Substances Hazardous to Health (COSHH) Regulations 2002
+ Provision and Use of Work Equipment Regulations (PUWER) 1998
+ Manual Handling Operations Regulations (MHOR) 1992
+ Personal Protective Equipment (PPE) at Work Regulations 1992
+ Work at Height Regulations 2005
+ Control of Noise at Work Regulations 2005
+ Management of Health and Safety at Work Regulations 1999
+ Construction (Design and Management) Regulations 2007
+ Environmental regulations
+ Waste management legislation.

Typical mistake

Many students misunderstand the difference between legislation, regulations and guidance. Make sure you learn the definition of each term and how they are distinct from each other.

Revision activity

Create a table with two columns. List as many regulations as you can in the first column, then try to recall the purpose of each in the second column. Check your answers with the regulations covered in this chapter.

Now test yourself   TESTED ◯

1 What is the role of the HSE?
2 What actions will an HSE inspector take if they believe that an employer has breached health and safety law?

1.2 Public liability and employers' liability

The implications of public and employers' liability

REVISED ◯

Public liability means employers have a legal responsibility to protect the public from injury, illness and death as a result of work activities.

Employers' liability refers to the responsibility of employers to protect their employees from harm in the workplace.

If a person is injured or suffers a loss in the workplace, they may seek financial compensation from the employer.

Under the Employers' Liability (Compulsory Insurance) Act 1969, all employers are required by law to insure against any liability for injury or disease to their employees.

There is no legal requirement for employers to have public liability insurance. However, it is recommended if the public are likely to be affected by work activities.

Legal action can be taken against an employer under public and employers' liability to cover any loss of income, medical costs and compensation.

Liability To have a legal responsibility for something

Typical mistake

Claims for compensation should be made by the injured person – they are not made by the HSE.

Revision activity

Create a table with two columns to show the costs of a workplace accident to an employer and to an injured person.

1.3 Approved construction codes of practice

The use, purpose and legal status of ACOPs

REVISED ●

The HSE publishes documents online that contain information and guidance for duty holders, with practical ways to comply with the law.

The HSE's Legal (L) Series publications (also referred to as the CDM Series) comprise Approved Codes of Practice (ACOPs) that describe preferred methods and standards. However, ACOPs only have a semi-legal status and, unless stated, they do not have to be followed. If another practical method is used, it must meet or exceed the standards in the ACOP.

Exam tip

You will not be expected to know all the ACOPs published by the HSE, but you should have a good understanding of those most relevant to onsite construction.

Typical mistake

There is no legal requirement for employers to follow ACOPs unless it is stated that they must do so. The legal status of HSE guidance and ACOPs is outlined on the Health and Safety Executive's website: www.hse.gov.uk/legislation/legal-status.htm.

Now test yourself

3　What is another name for the HSE's L Series?

TESTED ○

1.4 Implications of poor health and safety performance

Poor health and safety performance in the construction industry can have financial, legal and ethical consequences, including a negative impact on the environment. Table 1.1 summaries how poor health and safety impacts different individuals.

Table 1.1 Impacts of poor health and safety

Employee	Employer	Client/customer	Public
+ Accidents (such as slips, trips and falls) and near misses + Ill health (such as asbestosis and silicosis) + Injuries (e.g. from falls from height or onsite plant) + Death (e.g. electrocution) + Loss of work/income/quality of life + Recovery/treatment/rehabilitation	+ Lower productivity + Higher employee turnover + An unmotivated workforce + Financial problems, e.g. due to higher insurance premiums or legal costs + Damage to business reputation + Prosecution by the enforcement agencies, e.g. the Environment Agency and the HSE + Damage to work and equipment	+ Legal action taken by the enforcement agencies, e.g. The Environment Agency and the HSE, for failing to comply with their health and safety duties + Disputes with neighbours in the local community, e.g. due to noise, air and water pollution + Ill health + Death + Injuries + Damage to personal reputation + Increased costs	+ Accidents and near misses + Ill health + Injuries + Death + Environmental issues + Loss of work/income/quality of life + Damage to personal belongings or property

Typical mistake

Accidents do not just affect employers and their employees – they can also impact clients and customers, the general public and the business itself.

Addressing poor standards of health and safety in the workplace through control methods (for example, risk assessments, legislation) will reduce the number of injuries and fatalities, improve businesses reputation and performance, and reduce costs.

> **Risk assessment** Process used to identify, control and record hazards in the workplace

Health and Safety Executive powers of prosecution

If an employer does not follow health and safety legislation, the HSE may enforce HASAWA with its powers of prosecution, as outlined in section 1.1.

> **Revision activity**
>
> Research the HSE website for examples of cases when the powers of prosecution have been used.

1.5 Development of safe systems of work

Types of safe systems of work used in construction projects

The HSE favours that employers use the following approach:

+ Plan – for specific health and safety objectives
+ Do – implement the plan
+ Check – that the plan is working and measure performance
+ Act – learn from any mistakes and put them right.

The Management of Health and Safety at Work Regulations 1999 contain a schedule known as the 'General principles of prevention'. This provides a hierarchy of control measures to manage risks to health and safety in the workplace (see Figure 1.1).

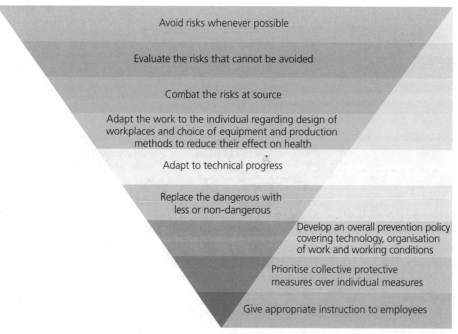

Figure 1.1 Hierarchy of control measures for managing health and safety risks (source: www.legislation.gov.uk)

Risk assessment

A risk assessment that identifies hazards and determines measures to eliminate or control them is fundamental to reducing work-related accidents and ill health. A risk assessment should be a site- and task-specific structured examination of workplace activities, appropriate and proportional to the level of risk and the nature of the hazards.

> **Hazard** Something with the potential to cause harm

Figure 1.2 Steps for completing a risk assessment

COSHH assessment

The steps for completing a COSHH assessment are as follows:

1 Identify the hazardous substance, who is likely to be harmed by it and how this may occur.
2 Evaluate the risk of the hazard causing harm by considering the frequency of exposure to the substance and what effects it could have.
3 Decide what measures are necessary to prevent or control exposure to the hazard and how these will be maintained, and plan for emergencies.
4 Record the assessment.
5 Decide if, and when, the assessment needs to be reviewed, and by whom.

Most effective method

Elimination: avoid or remove the hazard altogether

Substitution: replace hazardous substance with other, less hazardous substances

Engineering controls: isolate people from the hazard

Administrative controls: provide information, training and instructions to change the way people think and work, e.g. safety signs, site inductions and toolbox talks

PPE: as a last resort, protect individuals with personal protective equipment

Least effective method

Figure 1.3 Hierarchy of control measures to prevent harm from exposure to hazardous substances

Method statements

Method statements are documents prepared by employers that describe a logical sequence of steps to complete a work activity in a safe manner. A typical method statement describes:

+ hazards identified
+ safe access and egress (exit)
+ supervision needed
+ hazardous substances and how to control them
+ permit-to-work systems (if applicable)
+ personal protective equipment
+ emergency procedures
+ environmental controls
+ health and safety monitoring
+ workforce details.

How to apply CDM

Under the Construction (Design and Management) (CDM) Regulations 2015, principal contractors must engage with workers about their health, safety and welfare, and provide a site-specific induction and any other information and training they need.

> **Principal contractors** Contractors appointed by a client to take the lead in planning, managing, monitoring and co-ordinating health and safety in a project involving more than one contractor

Permit to work

Employers may adopt a permit-to-work system to manage high-risk activities on construction sites. This authorises people to carry out specific work tasks within a given timeframe and sets out the precautions required to complete the work safely.

Construction site signage

The Health and Safety (Safety Signs and Signals) Regulations 1996 state that safety signs should be used when:
+ there is a significant risk to health and safety that cannot be controlled in other ways
+ they can reduce a risk further.

See section 1.12 for categories of safety sign.

Certification schemes and qualifications

Construction Skills Certification Scheme (CSCS)

The Construction Skills Certification Scheme (CSCS) is accredited by the Construction Industry Training Board (CITB). CSCS cards prove that the card holder has a satisfactory level of health and safety awareness. They also show the card holder's relevant qualifications for their role on site.

Site Management Safety Training Scheme (SMSTS)

People with planning, organising, controlling and monitoring responsibilities are usually required by principal contractors and clients to hold this level of qualification.

Site Supervision Safety Training Scheme (SSSTS)

This qualification is designed for people with supervisory responsibilities or those preparing to start in this role.

> **Revision activity**
>
> Use a risk assessment template from the HSE's website to write a manual handling risk assessment for your place of work.

> **Now test yourself**
>
> 4 List the **five** steps for completing a risk assessment.
> 5 What is the purpose of a method statement?
>
> TESTED

1.6 Safety-conscious procedures

The benefits of safety-conscious procedures

Safety consciousness is an awareness of the presence of hazards and alertness to potential harm.

Safety-conscious procedures aim to promote and support safety consciousness within construction environments to keep people safe from harm.

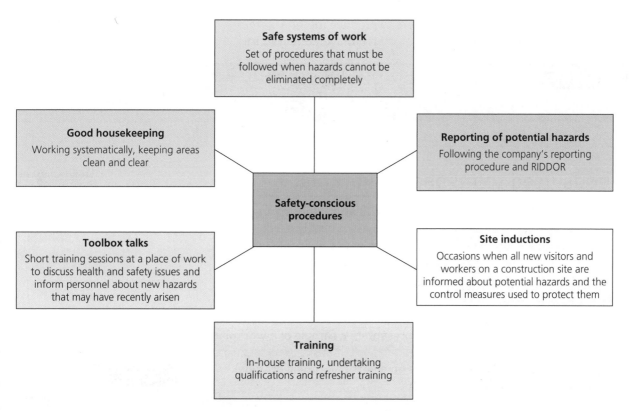

Figure 1.4 Safety-conscious procedures

The importance of safety-conscious procedures

REVISED

When legal responsibilities under HASAWA are not followed, duty holders are essentially breaking the law and could be putting themselves, and others, at risk of harm. Employers are at risk of prosecution by the HSE and compensation claims by the injured person.

If an accident or a near-miss incident occurs, it may also result in:
+ project timescales slipping
+ financial penalties due to missed deadlines
+ damaged company reputation
+ loss of business
+ difficulties retaining staff and recruiting
+ increased insurance premiums.

> **Exam tip**
>
> The examiner will expect you to know the difference between a hazard, a risk and a near-miss incident. You should also be able to list the hierarchy of control measures used to protect people from harm.

> **Typical mistake**
>
> A prohibition notice to close a construction site is not necessary every time a reportable accident occurs. For example, the HSE could issue a prohibition notice to prevent hazardous scaffolding being used until the faults have been rectified.

> **Now test yourself**
>
> 6 What is a toolbox talk?
>
> TESTED

1.7 Safety inspection of a work environment

Recording documents (risk assessments and method statements – RAMS)

REVISED

If an employer has five or more employees, risk assessments must be recorded in writing.

15

There is no legal requirement for employers to produce method statements, however this is recommended as part of a good management system.

> **Typical mistake**
>
> It is still recommended that employers complete risk assessments when they have fewer than five employees. However, there is no legal duty for them to be written down.

Methods used to inspect a workplace

REVISED

Employers have a responsibility to monitor health and safety arrangements in the workplace. Two types of monitoring system are typically used:

+ active monitoring – completed before an accident or incident occurs
+ reactive monitoring – completed after an incident has taken place.

Several types of health and safety inspection can be implemented in the workplace. These include:

+ health and safety audits (inspections of health and safety documentation)
+ safety sampling (used to focus on a representative sample of a workplace standard)
+ safety surveys (detailed investigations on a particular topic or issue)
+ safety tours (scheduled full inspections)
+ incident inspections (carried out after an accident, a near-miss or a case of ill health reported to the HSE)
+ visual or sensory inspections (unscheduled inspections of the work area, not restricted by a checklist or template).

Some regulations place specific duties on employers to review a work area, a process or resources, for example lifting equipment and PPE. The HSE provides information and guidance on these regulations. It also publishes documents that can be used to record the results of inspections, such as HSE forms F2534 and F2533.

> **Exam tip**
>
> The examiner will expect you to demonstrate a sufficient depth and breadth of understanding in your answers. You could show this by explaining the different types of health and safety inspection.

1.8 Recording and reporting of safety incidents and near misses

Recording and reporting

REVISED

Workplace accidents and incidents must be reported following the employer's reporting policies. This ensures that they are dealt with properly and investigated to reduce the risk of them reoccurring.

Employers must record the details of any accident in an accident book and keep accident records for at least three years.

> **Accident book** Formal document used to record details of accidents that occur in the workplace, whether to an employee or a visitor

RIDDOR puts duties on employers, the self-employed and people in control of work premises to report certain serious workplace accidents, occupational diseases and specified dangerous occurrences (near misses).

> **Typical mistake**
>
> Not all accidents need to be reported to the HSE – however, they must be logged in an accident book and records must be kept for three years.

> **Self-employed** State of working for oneself rather than an employer; a self-employed person is responsible for paying their own tax and National Insurance contributions on any earnings

> **Dangerous occurrences** Incidents that could have caused harm, injury or ill health

Check your understanding and progress at **www.hoddereducation.co.uk/myrevisionnotes**

Exam tip

Make sure you understand
the definition of dangerous
occurrences, so that you
can provide examples if
necessary.

1.9 Emergency procedures for unsafe situations

Unsafe situations   REVISED ○

Under the CDM Regulations, employers have a duty to plan for emergencies
on construction sites.

Under the Regulatory Reform (Fire Safety) Order 2005, employers also have
a duty to plan for emergencies on other sites such as offices, factories and
warehouses.

Lone working Employees
working by themselves
or without direct or close
supervision

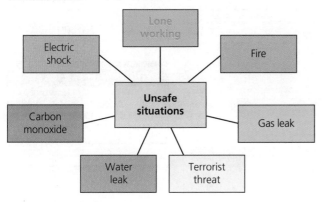

Figure 1.5 Examples of unsafe situations

Emergency procedures to follow if unsafe situations occur   REVISED ○

Gas leak

The Gas Industry Unsafe Situations Procedure (GIUSP) and Gas Safety
(Installation and Use) Regulations (GSIUR) 1998 state that the following
actions should be taken in the event of a gas leak:
+ Turn off the emergency control valve (ECV).
+ Open doors and windows.
+ Call the national gas emergency number.
+ Do not turn any power or light switches on or off.
+ Do not light any sort of flame.
+ Do not use any appliances.

Evacuations

Safe evacuation of the site should follow the designated emergency escape
route to the assembly point. A register of workers in attendance must be
completed.

17

Electric shock

Assess the situation, isolate the supply and call for help.

Injuries

+ Check that you and the injured person are not in any danger and, if possible, make the situation safe.
+ Call for help.
+ Carry out basic first aid.

Fire

In the event of a fire in the workplace, workers must follow their employer's procedures. The main steps are as follows:

1 Raise the alarm and inform others.
2 Walk quickly, following the directional signs, to the closest available emergency exit. Make sure you close all the fire doors behind you. Do not use any lifts between floors.
3 Only attempt to tackle a small fire if it is blocking your safe exit and if you are trained to use the equipment.
4 Report to the assembly point and stay there until you are told to leave.
5 Call the emergency services.

Fire extinguishers and their uses are summarised in Table 1.2.

Table 1.2 Fire extinguishers and their uses

Type of extinguisher	Colour of label	Fire classification
Water	Red	Class A
Dry powder	Blue	Class A
		Class B
		Class C
		Class D
		Electrical
Foam	Cream	Class A
		Class B
Carbon dioxide (CO$_2$)	Black	Class B
		Electrical
Wet chemical	Yellow	Class A
		Class F

> **Exam tip**
>
> There are several different types of regulations mentioned in this section – make sure you can distinguish between them.

> **Typical mistake**
>
> Foam extinguishers should not be used on electrical or Class F fires.

> **Now test yourself**
>
> 8 What classification of fires can a wet chemical extinguisher be used on?
>
> TESTED ○

1.10 Types of PPE

Purpose and correct use of PPE

REVISED ○

When the principles of prevention are applied to mitigate (reduce) the risk of harm, personal protective equipment (PPE) is always considered a last resort because it only protects the user.

Types of PPE

Table 1.3 summarises the types of PPE and the hazards they protect against.

Table 1.3 Types of PPE

Body part protected	Hazards	Types of PPE	Correct use
Ears	Noise	+ Ear defenders + Ear muffs + Ear plugs + Canal caps/semi-insert earplugs	Ear protection should reduce noise levels so that you are still able to communicate while wearing it. Ear protectors are manufactured with a single number rating (SNR) system, which allows the acoustic pressure on your ears to be calculated. Disposable foam ear plugs should be fully inserted in the ear to work properly and disposed of after each use.
Eyes	+ Sparks + Dust + Chemicals + Debris	+ Goggles + Safety spectacles + Face screens + Face shields + Full-face visors + Sunglasses	Eye protection should be: + compatible with other PPE worn + adjustable + stored correctly to prevent damaging the lenses.
Feet and legs	+ Slips + Falling objects + Objects (e.g. nails) penetrating the sole	+ Safety trainers, shoes, boots and Wellingtons with toecaps and protective mid-soles + Chainsaw and foundry boots + Knee pads + Kneeling pads	Footwear should: + have a good grip for different surfaces + be replaced when it becomes damaged. The risk assessment will identify which footwear should be worn.
Hands and arms	+ Cuts and abrasions + Impacts + Chemicals + Temperature extremes + Biological agents + Vibration	+ Anti-vibration gloves + Nitrile foam coated gloves + Gloves with cuffs + Gauntlets + Protective arm sleeves + Elbow pads	Care should be taken to select the correct type of gloves to protect against hazards. They must not create further risks, such as entanglement in machinery, when used.
Head and neck	+ Falling objects + Hair entanglement + Chemicals + Adverse weather	+ Hard hats + Bump caps + Snoods + Hair nets	Hard hats should be square on your head with the peak facing forwards. Avoid wearing caps or beanies underneath hard hats. Avoid marking hats with paint or pens (the chemicals may damage them). Bump caps should only be worn when there is a very low risk of bumping your head.
Lungs (respiratory system)	+ Dust + Vapours + Mists + Gases + Atmospheres with low or no oxygen	Respiratory protective equipment (RPE): + disposable half-mask respirators/dust masks + reusable half-mask respirators/dust masks with a filter + full-face mask respirators/dust masks + powered respirators with a mask/hood or helmet + breathing apparatus (BA)	Masks should form a good seal around the user's face to protect them properly. The type of masks and filters used should reflect the hazards. Employees should understand when and how to replace respirator filters. Masks should be stored correctly to prevent them being contaminated with hazardous substances.

Body part protected	Hazards	Types of PPE	Correct use
Whole body	+ Chemicals + Temperature extremes + Adverse weather + Dust + Metal splashes + Falling from height	+ Aprons + Overalls + Boiler suits + Chemical suits + High-visibility clothing + Harness	Whole-body protective equipment must be worn according to the manufacturer's instructions and should not cause a risk of entanglement with equipment or machinery. Contaminated PPE should be cleaned or disposed of properly and never mixed with personal clothing.

Typical mistake

Although employers are responsible for providing PPE free of charge, it is employees who are responsible for taking care of it and informing their employer when it needs to be replaced.

1.11 First-aid facilities

First-aid facilities in the work area

REVISED ◉

The Health and Safety (First Aid) Regulations 1981 place legal duties on all employers to provide adequate and appropriate first-aid equipment, facilities and people to assist their employees if they are injured or fall ill at work.

Employers must:
+ carry out a workplace-specific first-aid assessment to determine their needs
+ provide first-aid kits for their workers (including lone workers)
+ appoint a person to take charge of their first-aid arrangements and to call the emergency services when necessary
+ appoint a trained first-aider
+ provide staff training, information and instruction.

Exam tip

The examiner will expect you to understand the arrangements for first aid beyond a first-aid kit.

Typical mistake

Medicine should not be kept in a first-aid kit.

1.12 Warning signs for the main groups of hazardous substance

Categories of safety signs

REVISED ◉

The main categories of safety sign are listed in Table 1.4.

Table 1.4 Categories of safety sign

Type of safety sign	Description
Mandatory 	Tells you that something *must* be done, e.g. eye protection must be worn
Safe condition 	Shows directions to areas of safety and medical assistance in case of emergency

Type of safety sign	Description
Prohibition	Tells you that something *must not* be done, e.g. do not extinguish with water
Warning	Makes you aware of nearby danger, e.g. overhead load
Fire fighting	Marks the location of fire-fighting equipment and fire-alarm activation points

Typical mistake

Students often confuse prohibition and mandatory signs:
+ Prohibition signs are red and forbid certain types of behaviour, for example 'No access for unauthorised persons'.
+ Mandatory signs are blue and prescribe specific behaviour, for example 'Safety harness must be worn'.

CLP Regulations safety signs

REVISED

Manufacturers, importers, distributors and other users of chemicals have legal duties under the Classification, Labelling and Packaging (CLP) Regulations 2010 to use appropriate safety signs for labelling and packaging of hazardous substances and waste – as shown in Table 1.5.

Table 1.5 CLP Regulations safety signs

Safety sign	Meaning	Encountered when using ...
Explosive	Explosive, self-reactive	Gas
Flammable	+ Flammable gases, solids, liquids and aerosols + Self-heating, self-reactive + Contact with water creates flammable gas	+ Expanding foam + Nail-gun canisters + Solvent cement + Paint stripper

Safety sign	Meaning	Encountered when using ...
Oxidising	+ Oxidising gases, liquids and solids + May cause fire or explosion + May intensify fire	Chemicals
Gas under pressure	+ Contains gas under pressure + May explode if heated + Contains refrigerated gas which may cause cryogenic burns	Carbon-dioxide cylinders used in welding
Corrosive	+ Corrosive to metals + Causes severe skin burns and eye damage	+ Portland cement + Hydrated lime + Brick cleaner + Batteries
Acute toxicity	+ Toxic from single or multiple exposure + Toxic/fatal if swallowed, in contact with skin or inhaled	+ Materials containing formaldehyde + Hazardous air pollutants
Health hazard/hazardous to the ozone layer	+ May cause respiratory, eye or skin irritation + May cause drowsiness or dizziness + Harmful if swallowed, inhaled or in contact with skin + Harms the environment by destroying the ozone layer	+ Expanding foam + Grab adhesive + Wood adhesive + Solvent cement + Portland cement + Paint stripper
Hazardous to the environment	Toxic to the surrounding natural environment, especially aquatic life	+ Wood preservative + White spirit + Diesel, petrol and paraffin oils + Epoxy resin + Bitumen paint
Serious health hazard	+ May be fatal if swallowed or enters airways + May cause damage to organs + May damage fertility or cause genetic defects + May cause cancer + May cause allergy, asthma or breathing difficulties if inhaled	+ Expanding foam + Grab adhesive + Paint stripper + Wood dust + White spirit + Asphalt + Silica dust

Check your understanding and progress at **www.hoddereducation.co.uk/myrevisionnotes**

Exam tip

Focus on learning the different categories of safety sign rather than the different pictograms.

1.13 Safe practices and procedures for the use of access equipment and manual handling

Access equipment

REVISED ◯

If there is a risk of people falling any distance above or below ground that could result in injury, the employer must take the necessary precautions to eliminate the hazard completely or reduce the risk of harm to an acceptable level.

Where a risk remains, employers should use access equipment or other measures, such as safety nets, air bags or PPE, to minimise the distance and consequences of a fall.

Access equipment should only be used by trained, competent and authorised people in accordance with the manufacturer's instructions.

The safety aspects associated with different types of access equipment are summarised in Table 1.6.

Access equipment
Apparatus specifically designed for working safely at height

Ratio Relationship between two groups or amounts that expresses how much bigger one is than the other

Table 1.6 Safety aspects associated with different types of access equipment

Access equipment	Safety checks	Safe erection	Factors influencing choice of equipment
Ladder	Check the following parts have no visible defects before use: + rungs + stiles + anti-slip safety feet + guides + rung locks + locking mechanism. Check the ladder tag before use to ensure the equipment is safe to use.	Set at an angle of 75° or a ratio of 1:4. Place on firm, level ground. Place against a stable surface. Extend 1 m above a working platform. Secure to prevent slipping/moving.	Only use for light work and short durations. The user should always have three points of contact with the ladder and never overreach. The ladder must be secured to prevent slipping.
Mobile scaffold tower	Check the following parts have no visible defects before use: + frame + guardrails + toe boards + braces + platform + trap doors + castors and brakes + outriggers.	Place on firm, level ground. Erect in accordance with manufacturer's instructions. Do not overload. Use guardrails and toe boards. Apply brakes on castors before use. Correctly position and secure outriggers when needed to gain height. Do not reposition with people, materials or equipment on it.	It can be used by workers with both hands free. Care should be taken around overhead power cables while using or moving a tower. It is relatively quick to erect. It must be dismantled in certain conditions, e.g. high winds.

Access equipment	Safety checks	Safe erection	Factors influencing choice of equipment
Scaffolding	Safety inspections should be carried out after erection and before first use, and weekly thereafter. More frequent inspections will be needed after adverse weather. Scaffold tags (or 'scaftags') should be used to record the date of inspection and person who completed it. Check the following parts have no visible defects before use: + standards + ledgers + transoms + sole plates + handrails + intermediate rails + brick guards + toe boards + working platform/scaffold boards + braces + shoes + couplings and all other fittings.	Scaffolding must: + be designed and erected in accordance with British Standards and the Work at Height Regulations 2005 + have handrails 950 mm high, with no more than a 470 mm gap between guardrails + have toe boards 150 mm high + have platforms kept clean and clear.	Scaffolding must only be erected, inspected, adjusted and dismantled by trained and competent scaffolders. It is slow to erect and dismantle. It can provide a continuous working platform.
Trestles	Check the following parts have no visible defects before use: + toe boards and handrails + intermediate rails + steps/ladders + staging boards/scaffold boards.	Set up on firm, level ground. Erect in accordance with the manufacturer's instructions. Do not overload. Keep staging boards clean and clear. Ensure safe access and egress.	Trestles: + allow operators to work hands-free + are relatively quick to erect and dismantle + are suitable for low-height work.
Steps	Check the following parts have no visible defects before use: + steps or treads + prop + anti-slip safety feet + stiles + stepladder platform + locking mechanism. Check the stepladder tag before use to ensure it is safe to use.	Open the steps fully. Place on firm, level ground. Position facing the work, not sideways.	Only use for light work and short durations. They are quick and simple to erect and dismantle. The user must have three points of contact with the steps at the working position. The manufacturer's maximum safe working loads must not be exceeded.
Podiums	Check the following parts have no visible defects before use: + podium frame + locking mechanisms + elbows/hinges + platform + stabilisers + access ladder/steps + castors + anti-slip safety feet.	Set up on firm, level ground. Erect in accordance with the manufacturer's instructions. Keep the gate locked while working. Apply brakes on the castors before use.	Podiums are preferred to ladders and steps for long-duration work, as operators can work hands-free and have secure handrails. They are slower to set up, dismantle and move compared with ladders/steps.

Check your understanding and progress at **www.hoddereducation.co.uk/myrevisionnotes**

Access equipment	Safety checks	Safe erection	Factors influencing choice of equipment
Staging boards	Check the staging boards have no visible defects before use. They should also be clean and free from debris when in use.	Staging boards are used in conjunction with other types of access equipment, so they should be erected as instructed by the manufacturer, e.g. regarding minimum and maximum overhang.	Staging boards are sometimes preferred to scaffold boards with certain types of access equipment because they provide a wider platform without any trip hazards.
Boom and scissor lifts	Boom and scissor lifts must be set up and inspected ready for use in accordance with the manufacturer's instructions and LOLER Regulations.		Only suitably trained and competent people should use or operate boom or scissor lifts.

Access equipment should be regularly inspected for signs of wear and damage. Records should be kept of weekly, monthly and annual inspections. Some work equipment may be subject to specific requirements regarding inspection, such as the Lifting Operations and Lifting Equipment Regulations (LOLER) 1998.

> **Typical mistake**
>
> Falls from height can be above or below ground. Lanyards and safety nets will not prevent falls from height.

Manual handling

REVISED ○

Manual handling can be carried out by a single person, as a two-person lift, or using mechanical lifting aids.

Employers must take reasonably practicable measures to protect their employees from manual handling injuries. The Manual Handling Operations Regulations 1992 state that this should be done by:
+ avoiding manual handling if possible
+ assessing the hazards
+ reducing the risk as much as is reasonably practicable.

Employers have a duty to make sure that employees have the necessary information, instruction and training to perform manual handling operations. If it is not possible to avoid manual handling, good kinetic lifting techniques should be used.

> **Manual handling** Any lifting, carrying, supporting or moving of a load using bodily force
>
> **Kinetic lifting** Physical act of carrying, moving, lowering, pushing or lifting a load without the use of mechanical means

> **Exam tip**
>
> The best method of preventing musculoskeletal injuries from kinetic lifting is to avoid manual handling.

> **Now test yourself** TESTED ○
>
> 10 At what angle should ladders be placed when in use?
> 11 What is the best way to prevent manual handling injuries?

1.14 Safe practices and procedures for working in excavations and confined spaces

Dangers associated with excavations

REVISED ○

Excavations are often created on constructions sites to form trenches and holes for building foundations or to gain access to underground services and drainage.

Working in an excavation can be extremely dangerous because of the risks of:

+ flooding
+ collapse, causing people to be crushed
+ reduced levels of oxygen
+ poisonous gases (naturally occurring or created by work activities, for example exhaust fumes from nearby vehicles)
+ explosive atmospheres
+ coming into contact with buried services, for example gas pipes and electricity cables
+ unexploded ordnance
+ undermining nearby structures.

> **Unexploded ordnance**
> Explosive weapons that did not detonate when employed, for example Second World War bombs

Safe working in excavations

REVISED

Because of the associated risks, working in an excavation should be avoided and only considered as a last resort. The following control measures can be taken to reduce the risks:

+ Isolate the hazard, for example use edge protection.
+ Install stop blocks a safe distance from the excavation.
+ Put up safety signs.
+ Ensure adequate lighting.
+ Ensure safe access and egress.
+ Erect temporary support systems to prevent the excavation collapsing.
+ Ensure safe crossing points.
+ Carry out soil testing.
+ Implement a permit-to-work system.
+ Ensure adequate ventilation.
+ Identify and protect exposed services.
+ Carry out daily inspections.
+ Carry out atmospheric testing.
+ Use portable gas-detection equipment.
+ Ensure good systems of communication.

> **Stop blocks** Physical barriers used to prevent construction vehicles falling into an excavation

Dangers associated with confined spaces

REVISED

Additional hazards that may be found in confined spaces include:

+ fire
+ extremes of heat and cold
+ dust, fumes and vapours
+ flooding resulting in drowning
+ free-flowing solids (for example sand) causing suffocation
+ entrapment.

Hazards can also be created by workers with electrical equipment, machinery, materials and substances such as petrol.

> **Confined spaces**
> Workplaces that may be substantially but not always entirely enclosed, where there is a foreseeable serious risk of injury because of the conditions or from hazardous substances

Safety measures when working in confined spaces

REVISED

The Management of Health and Safety at Work Regulations 1999 state that:

+ employers and the self-employed must complete a risk assessment for work in a confined space
+ everyone involved in working in a confined space must be trained and competent
+ a permit-to-work system may be adopted to control work activities, for example communication from the inside to an outside sentry, how to raise the alarm, PPE, testing and monitoring, and a rescue plan.

> **Exam tip**
>
> Remember, a suitable and sufficient risk assessment must be completed before working in a confined space or excavation.

Typical mistake

A confined space is not always defined as small by its nature – so make sure you really know how a confined space is defined.

Exam tip

To gain higher marks, you must understand the differences between the command words 'state', 'identify' and 'describe' in exam questions.

Now test yourself

TESTED ◯

12 What term is used to describe a workplace that is substantially but not fully enclosed, with a foreseeable serious risk of injury because of the conditions or from hazardous substances?

13 What system is often used on construction sites to control high-risk activities?

Exam checklist

In this content area, you learned about the following:
+ construction legislation and regulations
+ public liability and employers' liability
+ approved construction codes of practice
+ implications of poor health and safety performance
+ development of safe systems of work
+ safety-conscious procedures
+ safety inspection of a work environment
+ recording and reporting of safety incidents and near misses
+ emergency procedures for unsafe situations
+ types of PPE
+ first-aid facilities
+ warning signs for the main groups of hazardous substance
+ safe practices and procedures for the use of access equipment and manual handling
+ safe practices and procedures for working in excavations and confined spaces.

Exam-style questions

Short-answer questions

1 State the primary legislation passed in 1974 to protect people in the workplace and the general public from work activities. [1]

2 List the hierarchy of control measures for dealing with the risks posed by manual handling. [3]

3 Explain the purpose of a risk assessment. [3]

4 Describe the implications of not adhering to health and safety legislation. [2]

5 Explain the purpose of the Construction Skills Certification Scheme (CSCS). [3]

6 Describe the key features of a toolbox talk. [2]

7 List the types of health and safety inspection that can be implemented in the workplace. [2]

8 Determine **four** implications for workers of poor health and safety. [4]

9 Explain the importance of keeping accident reports and records. [4]

10 Identify the **three** elements needed for a fire to burn. [3]

11 What responsibilities do employers and employees have with regards to PPE? [4]

12 State the legal responsibilities employers have towards their employees with regards to first-aid arrangements. [4]

13 A ladder is leaning against a scaffold platform 3.2 m high. Calculate the length of the ladder needed to safely reach the work platform. [2]

14 State the regulations that specify that access equipment must be suitable for its intended use, safe and well maintained. [1]

15 Describe how the risks of working in a confined space can be managed by an employer. [4]

16 Describe a prohibition safety sign. [2]

Extended-response questions

17 Evaluate why the Management of Health and Safety at Work Regulations 1999 describe the least effective control measure for employers to manage risks to health and safety in the workplace as giving appropriate instructions to employees. [12]

18 An apprentice bricklayer is employed by a contractor working on a small development project. The apprentice has some concerns about the standards of health, safety and welfare on the site, in particular with regards to the access equipment that he has been given to use. The apprentice has not received any training to use the equipment provided by his employer, and it seems to have some damage. The apprentice has reported his concerns to his site supervisor.

Identify the possible outcomes for the employer. [12]

19 A building contractor has six skilled workers employed full time. The contractor is aware of their responsibilities under the Management of Health and Safety at Work Regulations 1999 regarding written risk assessments. They have therefore consulted with their workforce and asked them to contribute to a risk assessment for power tools.

Analyse the risks of working with basic mains power tools and determine some control measures that could be implemented by the employer. [12]

2 Construction science principles

International System of Units (SI)

The International System of Units, commonly known as SI units, uses metric measurements. Base SI units are shown in Table 2.1.

Table 2.1 Base SI units

Quantity	Unit of measurement	Identification symbol
Mass	kilogram (kg)	m
Length	metre (m)	L
Time	second (s)	t
Temperature	kelvin (K)	T
Electric current	ampere (A)	I
Luminous intensity	candela (cd)	I_v

Derived SI units

Table 2.2 shows the derived SI units used in the construction industry.

Table 2.2 Derived SI units

Quantity	Unit of measurement	Identification symbol	Base formula (where relevant)
Area (or cross-sectional area)	Square metre (m²)	A	For squares or rectangles: length × width. For circles: $\pi \times$ radius² or πr^2
Volume	Cubic metre (m³)	V	For cuboids: length × width × depth or area × depth. For cylinders: $\pi r^2 \times$ depth or area × depth
Flow	Metres cubed per hour (m³/h) or commonly litres per second (l/s)	(mdot) $\dot{m}$	$\dfrac{\text{volume}}{\text{time}}$ or $\dfrac{\text{litres}}{\text{second}}$
Density	kg/m³	ρ	$\dfrac{\text{mass}}{\text{volume}}$
Velocity (speed)	Metres per second (m/s)	v	$\dfrac{\text{distance}}{\text{time}}$ or $\dfrac{\text{metres}}{\text{second}}$
Heat requirements using specific heat capacity	Kilojoules per kilogram per degree Celsius (kJ/kg/°C)	C_p	Heat required = volume × shc × temperature difference. Each substance has a set shc (specific heat capacity), e.g. the shc of water is 4.2kJ/kg/°C
Acceleration	Metres per second squared (m/s²)	a	$\dfrac{\text{velocity}}{\text{time}}$
Electromotive force	Volt (V)	ε	$\varepsilon = BLV$. Based on the length of conductor (L) in a field of magnetic flux density (B) and the velocity of movement (v)

Quantity	Unit of measurement	Identification symbol	Base formula (where relevant)
Electrical resistance	Ohm (Ω)	R	$$\frac{\rho L}{A}$$ Based on the material resistivity (ρ), length of the conductor (L) and the cross sectional area (A)
Illuminance (light on a surface)	Lux (lx)	E	Lumens per m2, where lumens are the measure of visible light emitted by a source
Internal pressure	Pascal (N/m2)	Pa	$$\frac{N}{m^2} \quad \text{or} \quad \frac{J}{m^3}$$
Atmospheric pressure	Bar	Bar(g)	Equal to 100 000 Pa
Energy (work)	Joule (J)	E	force × distance or $f \times d$ or power × time or watts × seconds
Force	Newtons (N)	F	mass × acceleration
Power	Watts (W)	P	$$\frac{energy}{time} \quad \text{or} \quad \frac{joules}{seconds}$$ In electrical circuits: volts × amperes or $V \times I$

Worked examples

1 Calculate the heat required to raise the temperature of 100 litres of water from 12°C to 62°C. The specific heat capacity of water is 4.2 kJ/kg/°C.

Heat required = volume × shc × temperature difference

= 100 × 4.2 × (62–12) = 21 000 kJ

2 Calculate the volume of a room measuring 2.3 m high, 5 m long and 6 m wide.

volume = length × width × depth

= 5 × 6 × 2.3 = 69 m3

Now test yourself

TESTED ◯

1 How do you calculate:
   a area?
   b volume?
2 What is the unit of measurement for force?

Atmospheric pressure
Force exerted on the Earth's surface by the weight of air above; this varies depending on height above sea level

Exam tip

Throughout your exam papers, you will be required to carry out calculations. Ensure you always show your working out.

Typical mistake

Students lose marks when they do not include the correct SI units in their answers.

2.1 Materials science principles

Materials and their properties

REVISED ◯

Metals

Different metals have different properties, which define how they are used in construction work:

✦ Iron is a ferrous metal with a high melting point that can be formed and shaped when heated. It is used for heat exchangers within boilers.
✦ Steel is a ferrous metal (technically an alloy) made from iron and carbon. It is rigid, strong and dense, so it is used for structural support and other instances where strength is required, such as rebar.

Ferrous metal Metal that contains iron

Rebar Reinforced steel bar commonly used in concrete to act as a frame to stop it moving and cracking

29

- Copper is a non-ferrous metal which is often used to create copper alloys. It is soft, rigid, durable, resistant to corrosion and anti-microbial, and is commonly used for pipework.
- Aluminium is a non-ferrous metal which is light but durable and can be easily formed. It is used widely in construction and is used for window frames, panels, roofing, guttering and door frames.
- Alloys are made by mixing different metals together to create a metal with improved properties, for example chromium is mixed with steel to produce stainless steel; the improved properties are high corrosion resistance and hardness.

Plastics

Thermosetting plastics and thermoplastics are soft, flexible plastics which can withstand high temperatures and are used to insulate cables.

Unplasticised polyvinyl chloride (uPVC) is rigid and used in guttering and window and door frames.

Rubber

Rubber is a flexible material that can be used in its natural state or manufactured to ensure it is durable and both heat and water resistant. It is used to insulate cables and for water seals.

Ceramics

Ceramics include a wide range of materials manufactured from minerals, to which metals are often added to provide increased strength. They are heat and corrosion resistant and are used in firebricks, tiles, worktops and as electrical insulators.

Aggregates

Aggregates are a mass made from loose or grains of materials. Examples include shingle or sand. These are commonly mixed with cement for mortar or concrete. Aggregates can also be used as standalone materials such as hardcore or gravel.

Lime

Lime is a white alkaline which is obtained by heating limestone. It is used in mortars as it retains water well, allowing less air in, which gives bond strength.

Cement

Cement is a mixture of lime and other materials such as clay and chalk. It is used as a binder in mortar or cement.

Mortar

Mortar is a mixture of cement and an aggregate, such as sand, with water to form a mixture used to bind together bricks or blocks when left to dry/cure.

The mixing ratios of cement are:
- 1:3 for exposed external use
- 1:4 for general external use
- 1:5 for sheltered internal use
- 1:7 for general internal use.

There are two ways to get the materials ratio correct before mixing:
- volume batching – where a container is used as a measure of the material
- weight batching – where the ratios are mixed by weight.

Copper alloys Metal alloys that contain copper and one or more other metals

Corrosion Gradual deterioration of metals through chemical or electrochemical reaction with their environment

Hardness Ability of a material to resist scratching, wear and tear, and indentation

Thermosetting plastics Plastics that once formed cannot be reformed

Aggregate Coarse mineral material such as crushed stone (gravel) used in making concrete

Hardcore Made from waste broken bricks, stone and/ or blocks; used to provide a base

Check your understanding and progress at **www.hoddereducation.co.uk/myrevisionnotes**

Concrete

Concrete is used to form solid objects such as bases, foundations or plinths. The mixing ratios depend on use. As a guide, the ratio of cement:fine aggregate:coarse aggregate is:
+ 1:2:4 for fence posts
+ 1:3:3 for floors, walls or footings
+ 1:2:3 for driveways
+ 1:2:2 for floors and foundations.

Site concrete testing

The quality of concrete must be monitored to ensure a uniform mix and consistency through different batches that are made, either onsite or delivered as ready-mixed.

Bricks

Bricks are generally made from clay-based materials, formed in moulds and fired hard in a kiln, and used to form walls. Bricks fall into three categories:
+ facing bricks
+ engineering bricks
+ common bricks.

The properties of a brick will depend on the type of brick. Manufacturer's data will provide information on crushing strength, water absorption, salt content and frost resistance.

Timber

Timber is used throughout the construction of buildings, mainly internally for structural supports for floors or internal stud wall partitions. Timber has two principal gradings:
+ GS for general structure
+ SS for special structure.

Sheet materials – wood based

Sheet materials are used in construction for a wide range of applications from flat roofs to partition walls, and come in various forms:
+ plywood: made from gluing together multiple layers of wood sheets
+ blockboard: softwood timber strips which are glued and sandwiched between two large sheets
+ chipboard or particle board: formed by pressing and gluing together small wood chippings or waste pieces of wood.

Plasterboard

Plasterboard is a sandwich of dried and hardened gypsum plaster between two sheets of paper. It is used extensively to provide a smooth internal wall or ceiling surface ready for decorating.

> **Gypsum** Natural mineral used in products such as cement, plaster and plasterboard

Paints

Paints can be categorised as water- or oil-based and these act as the binder, or main body of the paint. Pigments are added to the binder to give different colours or durability to the paint. Paint is generally used for two reasons:
+ appearance
+ surface/materials protection.

2.2 Mechanical science principles

Key principles of mechanical science

Mass and weight

Mass is a measure of the quantity of matter in an object. It is not dependent on gravity and is therefore different (but proportional) to weight.

Weight is the downwards gravitational force acting on a mass.

The relationship between mass and weight is defined by Newton's second law:

$F = ma$

where:
+ F = force (N)
+ m = mass (kg)
+ a = acceleration (m/s^2).

Newton's second law can express weight as a force by replacing acceleration (a) with acceleration due to gravity (g):

$W = mg$

where:
+ W = weight (N)
+ m = mass (kg)
+ g = acceleration due to gravity (on Earth, this is 9.81 m/s^2).

Work

Work is any force that lifts, pushes or twists an object resulting in movement.

When an object moves in the same direction as the force exerted, the work done is equal to the force exerted multiplied by the distance moved:

work = force × distance

When including the values used to determine force:

work = mass × gravity × distance

Mechanical work is measured in joules (J). Newton-metres (Nm) can be used for mechanical work, but are also used as a measurement for torque.

Other units used include the:
+ kilowatt hour (kWh) – used to measure electrical energy
+ calorie – used to measure food energy
+ BTU (British Thermal Unit) – used for heat-source applications, such as burning gas.

Energy

Energy is categorised into two main groups:
+ kinetic energy (the energy an object has due to its motion)
+ potential energy (stored energy held by an object).

The potential energy of gravity keeps a mass on the ground. If the mass is raised, then a machine uses kinetic energy. If the input of kinetic energy ceases, the potential energy tries to bring the mass back down to the ground.

Power

Power is the rate of doing work. It is calculated by dividing work done by the time taken to carry out that work:

$$\text{average power} = \frac{\text{work done}}{\text{time taken}}$$

The unit of power is joules per second (J/s), and 1 J/s is equivalent to 1 watt (W).

Check your understanding and progress at **www.hoddereducation.co.uk/myrevisionnotes**

Worked examples

1. What is the output (mechanical) power required for a motor to raise a mass of 2000 kg to a height of 6 m above the ground in one minute?

$$P = \frac{m \times g \times d}{t}$$
$$= \frac{2000 \times 9.81 \times 6}{60}$$
$$= 1962\,W$$

2. If the same motor raised the same load in 10 seconds, what is the output power required by the motor?

$$P = \frac{m \times g \times d}{t}$$
$$= \frac{2000 \times 9.81 \times 6}{10}$$
$$= 11772\,W$$

Note how the amount of energy used is the same, no matter how quickly the task is carried out – but more power is required to do the work in a shorter time.

Efficiency

The law of conservation of energy states that energy cannot be created or destroyed, only transformed from one form to another.

Some of the energy may be turned into unwanted forms, such as noise or heat, known as energy losses. This is common in mechanical processes.

The efficiency of a mechanical system is the ratio of output power compared to input power, expressed as a percentage:

$$\% \text{ efficiency} = \frac{\text{output power}}{\text{input power}} \times 100$$

Worked example

If a machine with a 100 kW output has an input power of 220 kW, what is the machine efficiency? Give your answer to one decimal place.

$$\% \text{ efficiency} = \frac{\text{output power}}{\text{input power}} \times 100$$
$$= \frac{100}{220} \times 100$$
$$= 45.5\%$$

Exam tip

Exam questions will tell you how many decimal places the answer should be given to. Ensure you follow the guidance to receive full marks.

Basic mechanics

REVISED

Theory of moments (torque)

Torque refers to how tight something needs to be, for example a bolt or a screw:

+ If there is too much torque (it is too tight), this may break the thread or snap the object.
+ If there is not enough torque (it is not tight enough), this may cause the object to come loose.

Torque, or moment, is measured in newton-metres (Nm) and calculated as:

$M = \text{force } (f) \times \text{distance } (d)$

Action and reaction

For every action, there is an equal and opposite reaction. For example, your weight pushes down on the floor and the floor pushes up against you with an equal force.

Centre of gravity

If a pivot is placed under an object's centre of gravity, the object remains balanced, as shown in Figure 2.1. This state is known as equilibrium.

If the pivot is placed anywhere else, the object is unstable and falls.

When lifting or manually handling, objects should be supported under the centre of gravity to make them easier to carry.

> **Centre of gravity**
> Imaginary point where the weight of an object is concentrated

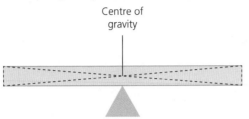

Figure 2.1 Finding the centre of gravity

Velocity and ratio

Velocity is the speed of something in a given direction, measured in metres per second (m/s).

Ratio is the relationship between two groups or amounts that expresses how much bigger one is than the other. For example, we might consider the ratio of the distance moved to the effort applied to the load.

Simple mechanics

REVISED ●

Levers

Levers exert a large force when applying a lesser force.

While levers give an advantage in terms of the force applied, the distance travelled is increased, meaning the energy used is essentially the same. This is because the torque applied to the ends of a lever must be equal but opposite. Therefore:

$$\text{force} \times \text{distance} = \text{force} \times \text{distance}$$

Torque is measured in newton-metres (Nm). This is the product of the force (N) applied to a lever and the distance (m) of the force from the fulcrum.

The advantage gained by a lever is called mechanical advantage (MA):

$$MA = \frac{\text{load}}{\text{effort}}$$

Levers are categorised into three classes, as shown in Table 2.3.

Table 2.3 Classes of lever

Lever	Description	Examples
Class 1	The force and the load are on different sides of the fulcrum. The greater the distance, the greater the effect of the force.	Pliers, catapult, scissors, seesaw, crowbar and claw hammer, when used to lift nails
Class 2	The load is between the force and the fulcrum.	Wheelbarrow, stapler and nail clipper
Class 3	The force is between the load and the fulcrum.	Fishing rod, tweezers, tongs, mousetrap and the human arm

Check your understanding and progress at **www.hoddereducation.co.uk/myrevisionnotes**

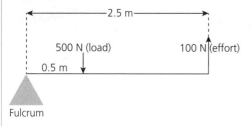
Calculate the mechanical advantage of the lever shown below:

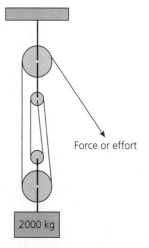

$$MA = \frac{load}{effort}$$
$$= \frac{500}{100}$$
$$= 5$$

2.5 m

500 N (load)   100 N (effort)

0.5 m

Fulcrum

When completing calculations ensure you show your working out and any units of measurement required.

Pulleys

A pulley gives a mechanical advantage when lifting an object. By running a rope through a four-pulley system, a mechanical advantage of four is gained. For each pulley you get one extra unit of mechanical advantage.

Calculate the force required to raise a mass of 2000 kg in the pulley system shown in Figure 2.2.

As the load has been described as a mass, first determine the downward force of the load:

force = mass × gravity

NB acceleration due to gravity on Earth = 9.81 m/s²

$f = 2000 \times 9.81 = 19620\,N$

As the system has four pulleys compared to one pulling rope, the mechanical advantage is 4:1.

$$MA = \frac{load}{effort}$$

Rearrange the formula for mechanical advantage to find the effort required:

$$effort = \frac{load}{MA}$$
$$= \frac{19620}{4} = 4905\,N$$

Therefore, a downward force (or effort) of 4905 N is required to raise the load.

Force or effort

2000 kg

Figure 2.2 A pulley system

Screws

A screw is a mechanism that converts rotational motion to linear motion. It consists of a cylindrical shaft with helical grooves or ridges called threads around the outside. Screws are commonly used as a fixing device.

Load bearing formulae

Load bearing calculations are used to determine the load bearing capacity of building components that support a structure. They are used to calculate the maximum load that can be safely supported. There are a number of calculations that can be completed to determine this. There are calculators available online to assist with these tasks.

Another calculation that is commonly used in the construction industry is soil bearing capacity. This is carried out to calculate the capacity of the soil to bear loads coming from a foundation. This varies depending on the type of soil.

3   If a machine with a 250 kW output has an input power of 300 kW, what is the machine efficiency?

TESTED

Using the internet, research the different methods for load bearing calculations for the following materials:
+ steel beams
+ concrete foundations
+ timber structural components.

2.3 Electricity principles

Generation

REVISED ●

Fossil fuels are burned to heat water, forming steam. This steam turns a turbine, which rotates a generator at high speeds to produce electricity.

> **Fossil fuels** Finite energy sources formed by the decomposition of organic matter beneath the Earth's surface over millions of years, for example coal, gas and oil

Table 2.4 Fossil fuels used in electricity generation

Type	Description
Gas	This is the most widely used fossil fuel. It can produce heat instantly, so creates electricity quickly.
Oil	This is used for regional generators, which create power at times of peak demand. The generators can deliver electricity immediately.
Coal	This was used widely in the past but it is highly polluting, so its use has been scaled down in the UK.
Nuclear	Nuclear energy is released from the core of atoms. This source of energy can be produced in two ways: ✦ fission – when atoms split into several parts ✦ fusion – when atoms fuse together.

Renewable fuel sources are described in section 5.9.

Transmission

REVISED ●

The National Grid is a network of mainly overground cables used to send electricity from generator stations. The transmission system uses a range of very high voltages:
+ 400 kV (known as the super grid)
+ 275 kV
+ 132 kV.

> **National Grid** Network of power lines supplying electrical energy around the UK

The towers used to carry electricity cables are known as pylons. The higher the voltage used in the transmission system, the bigger the pylons must be.

Pylons used for the 400 kV super grid have six cables suspended from them (three on each side). There is a single cable running between the tops of pylons, which acts as a common earth.

Transformation

REVISED ●

Electromagnetic induction is the process by which a current can be induced to flow due to a changing magnetic field.

Types of transformers

Transformers change voltages within the transmission system:
+ Step-up transformers increase the voltage.
+ Step-down transformers decrease the voltage.

> **Transformers** Devices used to change the voltage in one circuit to a different voltage in a second circuit

If voltage is increased, current will decrease; if voltage is decreased, current will increase.

All transmission voltages are three-phase, so the transformers need to be three-phase as well. This means they have three incoming wires into three windings, which step up or step down the voltage into three outgoing wires.

Where transmission systems need to be hidden, for example in urban areas, they may need to run underground. In this case, the voltage will be stepped down (for example to 132 kV) so it does not damage the insulation between the conductors in the underground cable.

Distribution

REVISED

Electricity is tapped off the National Grid to be distributed. These systems are known as distribution systems and they are looked after by distribution network operators (DNOs).

As distribution is much more localised, the voltages can be stepped down to lower values. Lower voltages mean cables in urban areas can be run underground. Underground systems are more expensive, so rural areas normally use cheaper overhead supplies.

In most cases, the underground cables in towns and cities are 11 kV. These supply the many substation transformers.

Why different voltages are required

REVISED

Different equipment requires different voltages:
+ 12 V is used for equipment such as downlights and central heating controls. These are connected to a 230 V supply and a transformer is used to reduce the voltage.
+ 110 V is used for power tools on a construction site. This voltage is much safer than 230 V.
+ 230 V is used within a domestic property for equipment such as televisions, microwaves and refrigerators.
+ 400 V is used for industrial equipment.

> **Domestic** Relating to a dwelling or home

Basic electrical circuit principles

REVISED

Electricity is the flow of electrons from one atom to another. Materials that are good conductors, such as copper, iron and steel, have electrons that move out of orbit from atom to atom.

When the material is connected to an electromotive force (emf) such as a battery, the flow can be controlled in one direction because the electrons are attracted to the positive plate of the battery. This is called charge.

Materials whose atoms keep their electrons in orbit make good insulators, such as rubber or PVC.

DC and AC circuits

In a direct current (DC) circuit, electrons flow in a single direction. In an alternating current (AC) circuit, the electric charge flow changes its direction periodically.

Table 2.5 Key electrical values and SI units of measurement

Electrical value	SI unit of measurement	Symbol
Charge	Coulomb (C)	Q
Current	Ampere (A)	I
Electromotive force or circuit voltage	Volt (V)	V
Resistivity	Ohm-metre (Ω-m)	ρ
Resistance	Ohm (Ω)	R
Power	Watts (W)	P

Ohm's law

Ohm's law explains the relationship between current, voltage and resistance in any electric circuit. It is applied to work out the quantities of a DC circuit and can be expressed as:

$V = IR$

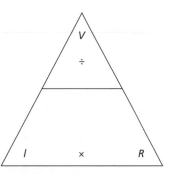

Figure 2.3 Ohm's law triangle

> ### Worked examples
>
> 1 Calculate the current in a 110V drill with 70Ω resistance. Give your answer to two decimal places.
>
> $$I = \frac{V}{R}$$
> $$= \frac{110}{70}$$
> $$= 1.57\,A$$
>
> 2 If a cooker has a 240V supply and a current of 30A, what is the resistance?
>
> $$R = \frac{V}{I}$$
> $$= \frac{240}{30}$$
> $$= 8\,Ω$$
>
> 3 Calculate the voltage in a circulating pump with a resistance of 44Ω and a current of 5A.
>
> $$V = IR$$
> $$= 5 \times 44$$
> $$= 220\,V$$

Series circuits

When resistors in a circuit are connected one after the other, they are connected in series. The total resistance is found by adding all the resistances together.

In a series circuit:
+ the current remains the same through each resistor
+ the voltage across each resistor is different.

The voltage across each resistor can be calculated by applying Ohm's law. The value of all the voltages must equal the total circuit voltage.

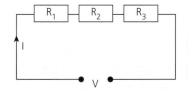

Figure 2.4 Circuit with three resistors in series

> ### Worked example
>
> Calculate the voltage across each resistor in the diagram below:
>
>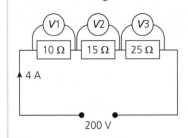
>
> Using Ohm's law:
>
> $$V = IR$$
> $$V1 = 4 \times 10 = 40\,V$$
> $$V2 = 4 \times 15 = 60\,V$$
> $$V3 = 4 \times 25 = 100\,V$$
>
> And to check: the value of all the voltages must equal the total circuit voltage:
> $$40 + 60 + 100 = 200\,V$$

Parallel circuits

When resistors are arranged in parallel, the supply voltage remains constant across all resistors but the current in each branch changes.

Measurement of electrical circuits

Instruments used to measure electrical quantities include the following:

+ A voltmeter measures potential difference, which is the voltage difference from one side of a load to the other. It is connected in parallel to the item being measured.
+ An ammeter measures small currents in series with the load.
+ An ohmmeter measures ohms but only works if the circuit or item is disconnected from the power source. It is connected in parallel to the item being measured.
+ A wattmeter measures the voltage and current and then calculates the resulting power. It connects both in parallel and in series.

Circuit protective devices

REVISED

Table 2.6 Types of protective devices

Protective device	Description
Fuse	This has a wire element that heats up with current: + If the current steadily reaches high values due to overloads, the wire melts gradually and the circuit disconnects. + If the current suddenly reaches high values due to a fault, the wire melts quickly and the circuit disconnects.
Circuit breaker	This has a magnetic coil. When a fault current reaches a pre-set value, the magnetic field rapidly trips a switch and disconnects the circuit.
Residual current device (RCD)	This has either an electronic device or a small toroidal transformer. It monitors the current entering a circuit through the live wire and the current returning through the neutral wire. + If the circuit is healthy, the two currents remain identical. + If a small fault happens, current flows to earth and the live wire has more current than the neutral wire. If the imbalance in current exceeds the device's residual current setting, it trips instantly. The most common residual current setting for an RCD is 30 mA or 0.03 A, so these devices are highly sensitive.
Residual circuit breaker with overload (RCBO)	This is a miniature circuit breaker and RCD in the same body.

Now test yourself

TESTED

4 List **three** non-renewable fuel sources.

5 What is the SI unit for power?

6 Calculate the current flowing through a 230 V circuit with a resistance of 115 Ω.

7 Calculate the total resistance for the circuit shown below:

Revision activity

For each circuit protective device, print an image of it and add notes to summarise its specific properties.

2.4 Structural science principles

Approved Document A of the Building Regulations 2010 requires a building to be constructed to withstand the combined dead, imposed and wind loads without deformation or movement that will affect stability.

Forces acting on a building

REVISED

Different types of force can act on a building:
+ Tension tries to stretch buildings or their components in opposite directions, for example cables in a suspension bridge. It can be resisted by using steel, which has good tensile strength.
+ Compression tries to crush a component by pushing on both ends, for example a column with the weight of the building at one end and the force acting on the ground below. It can be resisted by using materials that act well under compression, such as concrete and stone.
+ Shearing forces work in opposite directions to try to split or divide a component, for example wind blowing in different directions at different parts of a building. These can be resisted by using brick or concrete in walls.
+ Torsion tries to twist the component in opposite directions, for example buildings in strong winds. It can be resisted by using closed hollow sections or circular elements such as poles.
+ Bending acts on the centre of a beam that is supported at each end, for example wooden floor joists. It can be resisted by using reinforced concrete for long spans, or steel or wooden joists for shorter spans.

Material properties

REVISED

Material properties are terms used to describe how a material acts. They are summarised in Table 2.7.

Table 2.7 Material properties

Property	Description
Elasticity	Ability of a material to resume its normal shape after being stretched or squeezed
Plasticity	Ability of a material to undergo permanent deformation after being stretched or squeezed
Ductility	Ability of a material to withstand distortion
Durability	Ability of a material to withstand wear and tear
Fusibility	Ability of a material to be converted from a solid to a liquid using heat
Malleability	Ability of a metal to be worked without fracture
Temper	The degree of hardness in a metal
Tenacity	Ability of a material to resist being pulled apart
Conductivity	Ability of a material to conduct electricity

Structural members

REVISED

The supporting parts of a building are known as structural members. For a structure to remain sturdy, it is important that the thickness and strength of all members are calculated by structural engineers.

Approved Document A covers the loadings on a building, and the construction of the structural elements including the foundations, walls, floors, roofs and chimneys.

Loadings Application of a mechanical load or force on a structure

Check your understanding and progress at **www.hoddereducation.co.uk/myrevisionnotes**

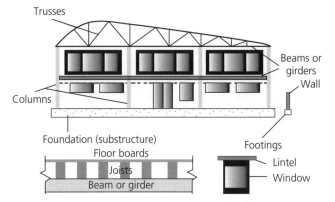

Figure 2.5 Structural members used in a building

Table 2.8 Structural members used in a building

Structural member	Description
Foundation	This is the lowest part of the substructure, supporting the building and preventing it from sinking into the ground. Metal or concrete piles may be required to make sure the building is on a firm substructure.
Substructure	This is the complete section of a building extending below ground-floor level.
Footings	These are the sections of masonry from the foundation to ground-floor level. They are normally linked to a particular member, such as a wall, not to the entire structure.
Columns	These are upright supports, which can be steel, concrete or brick. Brick columns are often referred to as pillars.
Beams or girders	These are horizontal elements that form the support for floors.
Joists	These are similar to beams but are often wooden and numerous. They are used to support a flooring system.
Trusses	These are normally associated with roofing systems and form triangles to provide support along large spans.
Lintels	These are used to provide structural support above windows and doors.
Superstructure	The portion of a building which is constructed above the ground level.

SWL stands for 'safe working load'. When you see a beam with 'SWL 500 kg' stamped on it, never suspend more than 500 kg from it. Loads acting on structures can be vertical, horizontal or longitudinal.

Drilling, notching and chasing

REVISED

Where services are installed in buildings, they need to be run through some structural parts.

Approved Document A outlines the requirements when drilling and notching joists and making holes in structural components, to maintain the integrity of the building:

+ The maximum diameter of holes should be 0.25 × joist depth.
+ Holes in the same joist must be at least three hole diameters apart.
+ The maximum depth of notch should be 0.125 × joist depth.
+ To avoid weakening the structure, vertical chases in walls should be no deeper than one third of the wall thickness, and horizontal chases should be no deeper than one sixth of the wall thickness.

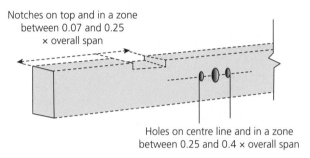

Notches on top and in a zone between 0.07 and 0.25 × overall span

Holes on centre line and in a zone between 0.25 and 0.4 × overall span

Figure 2.6 Notching and cutting holes in a joist or beam

Other effects on structures

REVISED

Various factors can affect a structure and weaken its stability:

+ Adjacent structures – air currents moving around adjacent structure can cause damage, as can soil movement caused by the foundations of other buildings.
+ Drains and sewers – underground systems can collapse if buildings are built on top of them, causing damage to foundations.
+ Trees – tree roots can damage building foundations.
+ Ground conditions – buildings can be damaged by subsidence (when the ground under a building collapses or sinks) or ground heave (when the soil beneath a structure expands and pushes the ground upwards).

A site survey will determine the adjacent structures and ground conditions for a building.

> **Now test yourself**
>
> 8 What is the maximum permitted diameter of a hole in a joist with a depth of 200 mm?
> 9 What is the maximum permitted notch depth in a joist with a depth of 220 mm?
>
> TESTED

2.5 Heat principles

Heat transfer

REVISED

Table 2.9 Types of heat transfer

Heat transfer	Description	Examples
Convection	This is the transfer of heat energy within a fluid or between a solid surface and a neighbouring fluid.	Radiators **Immersion heaters**
Conduction	This is the transfer of heat energy by direct contact between particles of matter.	Floor-heating systems Hot-water immersion systems
Radiation	This is the transfer of heat energy through empty space.	Outdoor patio heaters

> **Immersion heater**
> Electrical element that sits in a body of water; when switched on, the electrical current causes it to heat up, which in turn heats up the surrounding water

Characteristics of air

The characteristics of air are:
+ temperature – how hot or cold the air is
+ density – the mass per unit volume
+ humidity – the degree of moisture in the air.

Modern buildings strive to be more energy efficient by keeping heat in, but this can also cause problems:
+ If a building does not allow air changes to occur, air will become stale and contaminated, causing breathing problems.
+ If a building is allowed to heat up but moist warm air comes into contact with colder surfaces, such as windows, condensation forms, which in turn can cause mildew or rot.

What impacts heat loss in a building

Ventilation

Ventilation is air flowing through a building.

Approved Document F of the Building Regulations 2010 sets out the provisions for the ventilation of a building. When sizing emitters, the ventilation heat loss must be calculated using the following formula:

Ventilation heat loss = room volume × air change rate
× temperature difference × ventilation factor U-value of air (0.33 W/m³/K)

> **Emitters** Radiators or heaters used to heat a room

The temperature difference is calculated using the outside temperature and the base design temperature for a given room.

U-value is a thermal transmittance, or the heat loss through a structural element.

Worked example

Calculate ventilation heat loss for Room A where:
+ base design temperature = –1 °C
+ lounge design temperature = 21 °C
+ temperature difference = 22 °C
+ air change rate 1.5
+ room height = 2.3 m.

Give your answer to one decimal place.

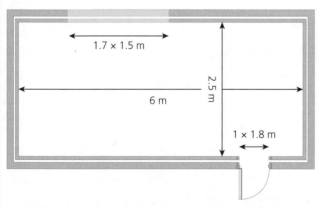

Ventilation heat loss = room volume × air change rate
× temperature difference × ventilation factor U-value of air (0.33 W/m³/K)

Room volume = 6 × 2.5 × 2.3 = 34.5 m³

Ventilation heat loss = 34.5 × 1.5 × 22 × 0.33 = 375.7 W

Building components

Heat loss in a building depends on the materials used in its construction.

Thermal transmittance is also known as the U-value. It is the measure of the overall heat loss through a wall or floor and is calculated considering the materials used, for example bricks, blocks and plaster, and their thicknesses.

Fabric heat loss = surface area × temperature difference × U-value

Heat loss occurs through the fabric of a room at a uniform rate through each surface. The U-values have been calculated for different types of construction, and the Chartered Institution of Building Services Engineers (CIBSE) guide provides U-value data for a range of specific building components. These are shown in Table 2.10.

Table 2.10 U-value data for different building components

Fabric	U-value (W/m³K)
Wall	0.35
Party wall	0.20
Ceiling	0.25
Internal wall (plasterboard)	1.72
Window	2.9
Floor	0.25
Door	2.9
Pitched roof (insulated)	0.16

The temperature difference is calculated using the outside temperature and the base design temperature for a given room.

Worked example

Calculate fabric heat loss through the walls for Room A where:

+ base design temperature = –1 °C
+ lounge design temperature = 21 °C
+ temperature difference = 22 °C.

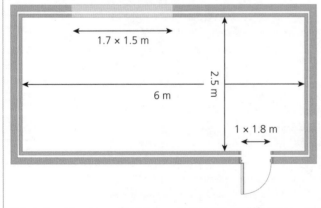

Fabric heat loss = surface area × temperature difference × U-value

surface area = 2.5 × 6.0 = 15 m²

Fabric heat loss = 15 × 22 × 0.35 = 115.5 W

2 Construction science principles

R-values

R-value is a measure of thermal resistance, or the ability of an object or a material to resist the flow of heat. This is calculated based on the thickness of the material:

$$\text{R-value} = \frac{\text{thickness of material (m)}}{\text{U-value (W/m}^3\text{ K)}}$$

Condensation

Condensation happens when warm, moist air comes into contact with cooler surfaces that are at or below the dew point. The most common places where condensation occurs are windows, walls and ceilings.

This can affect both the building fabric and the health of the occupants of the building.

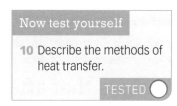

Now test yourself

10 Describe the methods of heat transfer.

TESTED ◯

2.6 Light principles

Light-emitting diode (LED) technology is used as the main form of lighting in buildings due to its energy-efficient performance and light quality. A range of other lighting options are also available.

Various factors are considered in the lighting of a building:
+ Refraction: the redirection of a wave as it passes from one medium to another.
+ Source of the light: there is a difference between light produced by electric lamps and light from the sun.
+ Glare: a bright light that is difficult to look at, which can be caused by poor positioning of luminaires.
+ Directed and reflected light: where the light cone is limited to a very small area.
+ Flow of light energy: the flow of light through free space or through a material medium.
+ Daylight factor: measure for the subjective daylight quality in a room.
+ Colour rendering: the appearance of light in terms of colour. Depending on the gases used in them, lamps produce either orange light or blue light. The different coloured light can change the appearance of items.

There are various different types of light that can be used:
+ General lighting service (GLS): a filament is suspended in a vacuum. When current flows through the filament, it glows white hot, producing light.
+ Tungsten halogen: a filament is suspended in halogen gas, which prevents evaporation. This allows the lamp to run at much higher temperatures than standard GLS lamps, meaning a brighter light.
+ High-pressure sodium (SON): current strikes across low-temperature sodium gas in a tube, causing it to heat up and ionise, producing light. The lamps are known as discharge lamps.
+ Low-pressure mercury: commonly known as fluorescent tubes, these mercury-filled tubes have an inner coating of phosphor powder, making a blue ultraviolet light inside the tube and various types of white light on the outside.
+ LED: made up of several light emitting diodes (LEDs), these lamps require control gear, known as drivers, to reduce and rectify voltage and govern current.

2.7 Acoustics principles

Acoustics in buildings

Acoustics is the study of mechanical waves that cause vibration and sound. Different materials and the shape of those materials affect the way sound waves move.

In construction, the aim is usually to provide acoustic comfort rather than enhance acoustics.

Poor acoustic protection can lead to hearing loss, high blood pressure, headaches and sleep deprivation.

Factors that affect acoustics

Factors that affect acoustics include:
+ frequency – the number of waves that occur in a defined period
+ reverberation – the persistence of sound following its creation, caused by sound reflecting off surfaces
+ reverberation time – the measurement (in seconds) of the time it takes for a sound to decay and stop
+ focusing – when sound is reflected from a surface at the same angle at which it strikes the surface
+ resonance – the frequency of sound and pressure which causes material to move easily and vibrate
+ echo – sounds caused by the reflection of sound waves from a surface back to the listener.

The unit of measurement for sound is decibels. A sound transmission class (STC) rating is a numerical value that indicates how well a structure reduces sound transmission. The higher the rating, the better the structure is at reducing sound travel.

Principles of sound

Soft materials, such as fabrics and foams, create a sound-absorbent surface where sound does not reverberate. Examples include:
+ cavity wall insulation, such as rock wool or fibre wool
+ isolation membrane (a thin barrier in walls, ceiling voids and doors)
+ flooring materials, such as carpets or wood
+ soundproof foam panels (usually shaped into small pyramids).

> **Sound-absorbent**
> Where sound waves are suppressed or absorbed by an item or a structure, rather than being reflected

Acoustic barriers are applied to the built environment to ensure privacy and control/limit unwanted transference of sound internally and externally. They include:
+ acoustic plasterboard
+ double glazing
+ insulation
+ sound-absorbing panels.

Laws and regulations on noise restriction

Laws restrict noise levels between 11 p.m. and 7 a.m. As a guide, constant underlying noise levels of 24 decibels (dBA) are permitted, with increases to 34 dBA intermittently.

Approved Document E of the Building Regulations 2010 sets out standards for new homes and conversions. The guidance is very detailed, including information on:
+ soundproofing, including the transmission of sound
+ prevention of unwanted sound travel within different areas of a building
+ the structure of materials and formation of elements of the building
+ the need to lag pipework or soundproof socket outlets recessed into a partition wall.

> **Now test yourself**
>
> 11 Define echo.
> 12 What is a sound transmission class (STC) rating?
>
> TESTED

Check your understanding and progress at **www.hoddereducation.co.uk/myrevisionnotes**

2.8 Earth science principles

Geology

REVISED

Ground structure

When constructing buildings, we need to consider the soil structure below the topsoil:

+ Low-rise buildings with shallow foundations typically have footings or foundations at least 1 m deep in clay-type subsoils, as these soils are prone to shrinkage or freezing due to their water content up to 0.75 m in depth.
+ Stony subsoils, where water is less likely to be held, may allow for shallower foundations, but this depends on the water table.
+ Where buildings require more stability, such as high-rise buildings, foundations need to be in contact with the bedrock layer. This is normally achieved using piles.

Greenfield sites are areas of land that haven't been developed or built on. Brownfield sites are areas of land that have previously been developed.

> **Topsoil** Upper layer of soil, usually between 50 and 200 mm deep, that contains most of the ground's nutrients and fertility
>
> **Water table** Top of an underground level where groundwater permanently saturates the land

Hydrology

REVISED

Hydrology is the study of water in the earth. A watercourse is a channel through which water flows.

Watercourses drain the land. Without them, water would cause soils to become unstable, leading to subsidence or even landslips.

When new developments are built, surface water needs to be drained into local watercourses, which can create flood risks if not managed correctly. Measures used to manage surface water, known as attenuation, include:

+ underground flood tanks that store water
+ the creation of ponds or reedbeds to store or absorb surface water
+ soakaways, which disperse surface water into the ground.

> **Hydrology** Study of water in the earth and its relationship with the environment

Earth forces and natural phenomenon

REVISED

Earthquakes

The Earth's crust is made up of large, moving pieces known as tectonic plates. Earthquakes occur when moving plates become jammed, creating a build-up of tension and forces. When the tension is released, an earthquake occurs.

The risk of earthquakes in the UK is low.

Landslides

A landslide is a mass movement of material, such as rock, earth or debris, down a slope. It can happen suddenly or more slowly over long periods.

Tidal factors

High and low tides are caused by the moon. The moon's gravitational pull generates something called the tidal force.

Weather

+ Wind can cause serious damage to buildings and structures.
+ Storms, persistent wind and rain can also damage buildings by causing erosion.
+ Flooding is becoming a more frequent occurrence due to climate change.

Climate change

Climate change refers to long-term shifts in temperatures and weather patterns. These shifts may be:

+ natural, such as through variations in the solar cycle
+ manmade, for example through rising atmospheric carbon levels due to the increased use of carbon-based fossil fuels.

> **Climate change** Large-scale, long-term change in the Earth's weather patterns and average temperatures

Now test yourself

TESTED ◯

13 Describe the difference between a brownfield and a greenfield site.

Exam checklist

In this content area, you learned about the following:
+ materials science principles
+ mechanical science principles
+ electricity principles
+ structural science principles
+ heat principles
+ light principles
+ acoustics principles
+ earth science principles.

Exam-style questions

Short-answer questions

1 State the SI unit for the following: [2]
   + mass
   + length.

2 State the mixing ratio for mortar for use externally in exposed areas. [1]

3 Describe the use of ceramics in construction projects. [1]

4 A mass of 2575 kg needs to be raised by a pulley with a mechanical advantage of six. How much effort is required? [3]

5 Which simple machine has a wheel on an axle that is used with ropes, chains, cables or belts? [1]

6 Describe the two methods of obtaining the correct ratio when mixing mortar. [2]

7 Describe the different types of paints. [2]

8 Describe the purpose of an acoustic barrier. [2]

9 Define atmospheric pressure. [1]

10 List **two** ferrous metals. [2]

11 State the figure used to represent acceleration due to gravity on Earth. [1]

12 State the three categories of bricks. [3]

13 Describe the working principles of a screw. [3]

14 Explain the purpose of Approved Document A. [6]

15 Describe building subsidence. [3]

16 Explain what timber is used for in construction projects. [1]

Extended-response questions

17 Describe the use of concrete and mixing requirements for construction projects. [9]

18 Explain **three** types of heat transfer. [12]

3 Construction design principles

3.1 Benefits of good design

The benefits of good design

REVISED ⬤

The benefits of good building design are summarised in Figure 3.1 and detailed below.

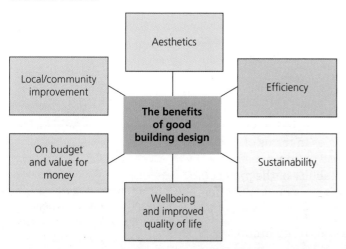

Figure 3.1 The benefits of good building design

All construction work in the UK must conform to building regulations and industry standards, however it may also be subject to planning permission, local restrictions and environmental requirements.

Aesthetics

Aesthetics refers to the appearance of a building in terms of style, materials used and finish.

Efficiency

Energy-efficient buildings use less energy than conventional buildings and consequently have lower running costs.

Sustainability

Sustainable buildings (also known as green buildings) have a low impact on the environment because of the use of infinite, recycled and reused building materials. They also use locally sourced materials to reduce the carbon emissions produced through transportation.

Sustainably built houses are less dependent on fossil fuels and often use renewables as a source of energy.

Wellbeing and improved quality of life

The arrangement, positioning and proportion of fenestration in the façade of a building will determine the amount of natural light and ventilation that enter it. This can give the occupants a feeling of space.

The direction a building faces will determine the extent and timing of the sunlight on the outside areas most often used.

Building regulations
Mandatory building standards in the UK

Planning permission
Approval that must be granted by the local authority for certain types of construction work

Carbon emissions Carbon dioxide released into the atmosphere, which is a cause of climate change

Renewables Natural sources of energy, for example wind, tidal and solar

Fenestration Openings in the façade of a building, for example windows and doors

---

On budget and value for money

Building work must stay within the client's budget. Therefore, design must:

+ not be unnecessarily complex
+ have an efficient method of construction
+ take into account the cost of specially designed and manufactured building components.

Local/community improvement

This concerns enhanced infrastructure, local facilities and affordable housing.

Infrastructure is the basic systems and services that a country or organisation needs in order to function properly, including transportation systems, water supplies, sewers and telecommunications systems.

Affordable housing is housing for sale or rent, for those whose needs are not met by the market (including housing that provides a subsidised route to home ownership and/or is for essential local workers).

> **infrastructure** Basic systems and services required for the proper functioning of society

The implications of poor design

REVISED ●

A project with poor design often has reduced efficiencies in terms of materials, labour and other resources that will increase building costs. The higher cost of construction either decreases the profitability of the project or increases the sale costs. Buildings that are poorly designed and do not offer value for money have reduced saleability.

Badly designed buildings will have a negative impact on the community and its environment. Projects that are built on greenfield sites damage the environment's natural habitat of flora and fauna.

The construction of new homes leads to increased traffic and puts a strain on current public transport systems.

The residents of new homes will increase the local population. Existing schools, healthcare facilities, shops and the job market may not have the capacity for expansion, therefore these are factors that must be considered at the design stage of a project.

> **Greenfield sites** Areas of land that have not been previously developed or built on, above or below ground (sites that have previously been developed are known as brownfield sites)
>
> **Flora** Plants and trees in a particular region
>
> **Fauna** Animals in a particular region

Factors that can impact on the profitability of projects

REVISED ●

Vernacular construction

The design of houses in a development may have to be sympathetic or particular to the region, relying on locally sourced materials and traditional skills. This method of region-specific house building is referred to as vernacular construction and is slower and more expensive.

> **Vernacular construction** Construction methods sympathetic or particular to a region
>
> **Energy performance certificate (EPC)** Document that provides an energy-efficiency rating for a building

> **Typical mistake**
>
> Vernacular construction has nothing to do with the use of construction plant. It is a term used to describe construction methods sympathetic or particular to a region.

Codes for sustainable homes

The design, construction and use of new homes must meet the demands of:

+ an environmental impact assessment (EIA)
+ an energy performance certificate (EPC)
+ a site waste management plan (SWMP)
+ Building Regulations (Approved Document L).

Check your understanding and progress at **www.hoddereducation.co.uk/myrevisionnotes**

Corporate social responsibility (CSR)

Construction businesses have moral responsibilities to protect the world around them. Corporate social responsibility (CSR) refers to strategies adopted by businesses to ensure they monitor and manage their social, economic and environmental impact on all aspects of society.

Brownfield vs greenfield sites

Constructing on a brownfield site protects greenfield sites. However, developing on these sites is usually more expensive because:
+ they may contain contaminated soil from previous use
+ existing structures or buildings may have to be removed from above or below ground
+ existing services to the site may be unsuitable and need to be upgraded or replaced.

Project sales

Poor design can reduce the saleability of new homes. If sales are slow in the initial phases of a project, this may prevent further phases being built due to the strain on project finances, for example repayments to investors.

Over-specification

Construction materials that are over-specified lead to higher build costs. For example, timber joists used to construct a suspended floor should conform to building regulations Approved Document A: Structure. However, specifying the stress grading and section of these materials beyond the requirements of building regulations would unnecessarily increase the total project costs.

Difficulty of assembly

Unfamiliar, complex designs or construction methods are often more difficult to assemble, leading to reduced efficiencies and increased timescales and budgets.

> **Corporate social responsibility (CSR)**
> The commitment of an organisation to carry out its business activities in a socially and environmentally responsible way

> **Exam tip**
> Make sure you understand the difference between building regulations and planning permission.

> **Now test yourself** TESTED ⚪
> 1 What term is used to describe buildings that have a low impact on the environment?
> 2 What does the term 'vernacular construction' mean?
> 3 Who is responsible for granting planning permission?

> **Revision activity**
> Write a short statement explaining why greenfield sites should be protected.

3.2 Design principles

Key definitions REVISED ⚪

Effective building designs are influenced by the design principles outlined in Table 3.1.

Table 3.1 Design principles – key definitions

Design principle	Definition
Environmental protection	Good building design should protect the local and natural environment The integration of sustainable technologies, energy-saving materials and alternative energy sources into a building design will also lower its carbon footprint.
Safety	A designer must consider how a building is going to be constructed, serviced and maintained in the future to protect the health and safety of construction and maintenance operatives.

Design principle	Definition
Speed and economics	New methods of construction are being developed to improve the efficiency of work on site, reduce build costs and increase profit margins.
	Off-site construction is often used for certain types of buildings, e.g. timber-frame and modular construction.
	Automation is used for some tasks that were once completed manually.
Aesthetics	The appearance of a building (e.g. design features, colour or materials used) may be influenced by modern or traditional design and functionality, or protected by listed and heritage building regulations.
Buildability manufacture	Prefabricating elements of a building, or whole modules, in a factory off site may be advantageous because it: + is a quick method of construction + does not delay production because of the weather + reduces onsite build times + reduces labour requirements/costs + has better quality control + reduces health and safety risks on site.
Installation and construction feasibility	Permission may have to be granted by the local planning authority before any construction work can take place. The site may be in a conservation area or national park. Sites may contain an existing building or structure that may be subject to listed and heritage building regulations.
Provision/integration of services	There will need to be access to existing mains services on site, or alternative services such as ground/air source heat pumps, bore holes for water or a sewage-treatment plant for waste.
Infrastructure	This refers to the development of new and existing roads, car parks, trainlines/stations, footpaths, bridges, bus stops, tunnels, schools, health centres, etc., both on and off site.
Inclusivity	This means building to meet the needs of the whole community, e.g. affordable homes and homes for young families, those with disabilities and elderly people.
Accessibility	There need to be good links to the site via footpaths, roads or rail.
Heat	As well as considering how heat is generated, it is important to protect a building from heat loss by using: + insulation in the walls, floors and roof + energy-efficient windows and doors.
Acoustics	This concerns the protection of the building and its occupants from unwanted sounds, e.g. from traffic.
Lighting and air quality	There needs to be: + access to natural light though the careful positioning of the fenestration in a building + ventilation, whereby stale air is removed from the building and replaced with natural fresh air.

Modular construction Combining factory-produced, pre-engineered units (or modules) to form major elements of a structure

Listed and heritage building regulations Formal guidelines that exist to protect and preserve buildings of special architectural or historical interest

Exam tip

The examiner will be looking at the depth of your understanding about good design principles and the effects of poor design.

Table 3.2 provides a comparison of traditional and modern construction methods.

Table 3.2 Comparison of traditional and modern construction methods

Method of construction	Traditional/ modern	Advantages	Disadvantages
Brick and block	Traditional	+ Familiar + Good thermal performance + Good sound insulation + Can be used with concrete upper floors	+ Thick walls due to insulation between inside and outside skin + Takes time to dry out + Cannot be carried out in heavy rain or freezing conditions

Check your understanding and progress at **www.hoddereducation.co.uk/myrevisionnotes**

Method of construction	Traditional/ modern	Advantages	Disadvantages
Open panel – timber frame	Traditional	+ Sustainable building material if the timber is from a managed forest + Factory-built framework reduces onsite build times	+ Can be more expensive than other building methods + Water can stain the exposed timber in the early stages of construction
Straw-bale construction (non-structural/infill system)	Traditional	+ Sustainable + Low cost + Quick to build	+ At risk of damage from vermin + Increased risk of fire + Uneven wall-surface finishes
Panelised – timber frame	Modern	+ Manufactured off site in a factory, thereby reducing onsite build times, labour requirements/costs and health and safety risks (e.g. working at height) + Quick to erect the shell of the building to make it watertight	+ Unable to support concrete upper floors + Long **lead time** for panels to be made off site + Liable to rot if exposed to moisture
Insulated concrete formwork (ICF)	Modern	+ Excellent thermal performance + Good sound insulation + Good weather exclusion	+ Inexperienced workers may need training to use the system
Steel-frame construction	Traditional	+ Quick to erect + Excellent weather resistance + Durable	+ Expensive + Heavy
Thin-joint blockwork/ masonry	Modern	+ No drying out of mortar joints needed + Quick to build + Excellent weather exclusion	+ Slightly more expensive compared to traditional bricks and blocks
Structural insulated panels (SIPs)	Modern	+ Good thermal performance + Manufactured off site in a factory, thereby reducing onsite build times, labour requirements/costs and health and safety risks (e.g. working at height)	+ Precise foundations required to align the SIPs; any deviation could result in costly delays + Requires a crane to lift the panels into position
Volumetric (pod/ modular)	Modern	+ Manufactured off site in a factory, thereby reducing onsite build times, labour requirements/costs and health and safety risks	+ Initially more expensive + Requires a crane to move the pods into position, which can be expensive to acquire if not already on site

Skin Single thickness masonry wall

Lead time Period of time between ordering and receiving goods or materials

The Royal Institute of British Architects (RIBA) Plan of Work

The RIBA Plan of Work is a design and process management tool used to bring greater clarity for the client at different stages of a project.

It organises the process of briefing, designing, constructing and operating building projects into the eight stages outlined in Table 3.3. Each stage has intended outcomes, core tasks and information that should be exchanged with different parties.

Table 3.3 RIBA Plan of Work: Stages and Stage Outcomes

	Stage	Outcome
Pre-design	0 Strategic definition	Determine the best way of achieving the client's requirements and the most appropriate solution.
	1 Preparation and briefing	Develop the client's concept and make sure it can be accommodated on site. Make sure everything needed for the next stage is in place.
Design	2 Concept design	Make sure the look and feel of the building is meeting the client's expectations and budget.
	3 Spatial co-ordination	Design the spaces within the structure of the building, before preparing detailed information about manufacturing and construction.
	4 Technical design	Develop information received from the design team and specialist subcontractors for the manufacture and construction of the building.
Construction	5 Manufacturing and construction	Manufacture and construct the building.
Handover	6 Handover	Complete the building works and address any defects that have been identified, to conclude the building contract between the client and the contractor.
In use	7 Use	The building should be used, operated and maintained efficiently until the end of its life. At this stage, the client may consider appointing professionals for aftercare activities such as servicing and maintenance.

Source: RIBA Plan of Work 2020 Stages and Stage Outcomes, reproduced courtesy of the Royal Institute of British Architects

Now test yourself

 TESTED ⬤

4 What is prefabricating elements of a building, or whole modules, in a factory off site known as?

5 What formal guidelines exist to protect and preserve buildings of special architectural or historical interest?

Revision activity

Create a mind map for the design principles for a building, using different coloured pens for different elements. Without looking, try to recreate the mind map the following day to see how much you can recall.

Typical mistake

The purpose and order of the stages in the RIBA Plan of Work are often misunderstood. Remember: it is a design and process management tool used to bring greater clarity for the client at different stages of a project.

3.3 Role of different disciplines involved in design

Key job roles

REVISED ⬤

Key job roles involved in construction design are outlined in Table 3.4.

Table 3.4 Key job roles involved in construction design

Job role	Key activities/responsibilities	Reporting lines/ lines of escalation	Potential career progression routes
Architect or other professional occupation	Design new buildings or structures Conserve or redevelop old buildings or structures	Client, the local authority and principal contractor	Lead/senior architect or project manager

Check your understanding and progress at **www.hoddereducation.co.uk/myrevisionnotes**

Job role	Key activities/responsibilities	Reporting lines/ lines of escalation	Potential career progression routes
Planner	Safeguards the land in villages, towns, cities, the countryside and commercial sites so it is used effectively to meet economic, social and environment needs	Consultants, clients and principal contractors	Independent planning advisor
Building inspector (**Local Authority Building Control, LABC**)	Inspects plans submitted for full planning approval, making sure they meet building regulations Conducts site visits to check compliance with regulatory standards of design and safety	Contractors and clients	Private building inspector
Quantity surveyor	Studies building information and drawings/ **Building Information Modelling (BIM)** to prepare tender packages Controls budgets and costs Provides professional advice to the client on quantities, time scales and quality standards Assigns work to subcontractors	Planning team or clients	Senior or chartered quantity surveyor
Land surveyor	Plots, measures and gathers/analyses data on land using GPS, surveying instruments and satellite images for construction and civil engineering projects Advises the client on planning and construction	Planning team or clients	Senior land surveyor
Building surveyor	Undertakes onsite property surveys Identifies defects and makes recommendations for remedial work Completes reports and advises the client on legal, planning or environmental issues	Planning team or clients	Senior building surveyor
Civil engineer	Designs, plans and supervises the construction and maintenance of public infrastructure, such as roads, railways, bridges and tunnels	Clients, architects, and consultants	Chartered, senior or master civil engineer
Draftsperson	Creates and modifies technical drawings using computer-assisted design (CAD) Visits sites to co-ordinate with architects, engineers and building services teams	Engineers, architects, construction site managers, clients	CAD designer or architectural technician
Clerk of works	Checks drawings and specifications, to ensure work is completed to the standards agreed in the building contract	Clients, site managers	Senior, chartered or master clerk of works
Manufacturer	Responsible for the design, production and sale of building materials and goods. Before products are sold and distributed, they are tested and certified to ensure they adhere to industry standards and regulations	Architects, engineers, designers, buyers, building contractors, client	Managerial, supervisory or advisory role

Local Authority Building Control (LABC) Local authority department responsible for inspecting building work against building regulations and signing off completed projects

Building Information Modelling (BIM) Use of digital technology to share construction documentation and provide a platform for collaboration

Exam tip

Underpin your answers about the disciplines involved in design by describing their key activities/responsibilities.

Now test yourself TESTED ◯

6 Explain the role of a mechanical building services (design) engineer.

7 When should a principal contractor be appointed for a construction project?

3.4 Design process from conception to completion

Design process

REVISED ◯

No two construction sites are ever the same. It is therefore essential that the land is properly investigated to identify any potential issues early and reduce the risk of additional unforeseen delays and costs at a later stage.

The key stages of the design process are summarised in Figure 3.2.

Initial enquiry

Most construction projects start with a client's needs. The client will often approach a designer to translate their initial ideas into a feasible design while keeping within budget.

Research

Before construction work begins, a desktop survey is usually undertaken to identify and record details of previous and current uses of the site. This will include:
+ site history
+ waste records
+ geology and hydrology
+ contamination reports
+ site boundaries
+ position of existing services
+ existing structures or buildings
+ local roads
+ access to the site
+ topography of the land
+ hedges, trees and fences
+ wildlife and habitats.

If a site of special scientific interest (SSSI) is identified, it must be protected from construction activities.

Site analysis

After the desktop survey has been completed, a walkover survey is usually undertaken. This is a physical inspection of a building site to identify any geological, ecological or topographical issues that may impact the project.

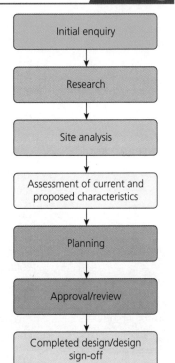

Figure 3.2 Key stages of the design process

Desktop survey Administrative investigation into a piece of land, completed without visiting the site

Topography Physical features and shape of land surfaces

Walkover survey Physical inspection of a building site

Analysis of the site will include looking at its size and geographical location, its topography and how the building project will impact on the land and its surroundings.

An inspection of a site may involve drilling boreholes into the ground at various positions to investigate the:
+ composition of the soil
+ load-bearing capacity of the ground
+ position of the water table.

The information gathered from the walkover survey is interpreted and used to determine the most appropriate foundation.

Assessment of current and proposed characteristics

During this stage, the design team:
+ analyses the current and proposed characteristics of the site
+ considers how these will impact or be integrated into the project design.

Planning

Under permitted development rights, you can extend or make certain alterations to an existing property without planning permission from the local authority. However, if the proposed changes are beyond the specific limitations of the rules, or new dwellings are to be constructed, then planning permission must be sought and granted.

To improve the efficiency and cost of a full planning application, a pre-planning application can be made.

If planning permission is required, an application must be made to the local authority by the client or planning team.

Minor construction work can be carried out on domestic buildings by submitting a building notice to the local planning department. Although work can start immediately once it has been submitted, lack of detail in the notice could result in work not complying with building regulations and therefore having to be dismantled or corrected after it has been completed.

Building regulations approval may also be required for proposed building work.

If a building is listed, then it is the responsibility of its owner to make a listed building consent application to the local authority before demolishing, altering or extending it.

> **Full planning application** Detailed formal request to a local authority planning department prior to building work taking place
>
> **Building notice** Basic application form sometimes sent to a local authority planning department to inform it of the intention to complete minor building work

Approval/review

There are three possible outcomes for a planning application:
+ approved
+ approved with conditions that must be complied with
+ refused.

An appeal can be made if an application has been refused.

Once building work has commenced, it must be inspected at regular intervals by either a building control officer (BCO) from Local Authority Building Control (LABC) or a private building inspector.

Completed design/design sign-off

Once planning permission is granted, the building design can be signed off and the construction phase can commence.

As soon as possible after a building project has finished, the contractor or client must notify LABC for a final inspection. Once LABC is satisfied with the building work, it will issue a completion certificate.

> **Exam tip**
>
> When answering an extended-response question about obtaining planning permission, you must be clear about the possible outcomes of an application.

57

Factors that may impact or influence design changes

Feasibility study

A feasibility study of a proposed construction project often identifies factors that impact or influence design changes, for example:

+ end-user requirements
+ budget
+ animals
+ infestation of animals or pests
+ sites of special scientific interest (SSSIs)
+ tree preservation orders (TPOs)
+ planning for utilities and connection to services (gas, water, electric, drainage)
+ frontage line
+ building line.

Frontage line and building line may be stipulated by the local planning department as a condition of planning approval.

> **Tree preservation orders (TPOs)** Legal protection of trees from damage, cutting, uprooting or removal
>
> **Frontage line** Front part of a building that faces a road
>
> **Building line** Boundary line set by the local authority beyond which building work must not extend

Construction (Design and Management) (CDM) Regulations

Individuals and organisations have legal duties under the Construction (Design and Management) (CDM) Regulations 2015. Designers must:

+ ensure the client is aware of their CDM duties and help them to comply
+ consider any preconstruction information provided by the client
+ eliminate foreseeable health and safety risks where possible
+ seek to reduce or control any health and safety risks that cannot be eliminated
+ provide design information for inclusion in the health and safety file for the project
+ co-ordinate with other designers working on the project on matters of health and safety during the construction phase and beyond
+ co-ordinate, communicate and co-operate with all contractors working on the project, taking into account their knowledge and experience of building design.

Project planning

Before the procurement of any construction resources, the project planning team will estimate building costs based on the information contained in the following documents:

+ working drawings
+ specifications
+ bill of quantities
+ take-off sheets
+ contracts
+ schedule of rates
+ estimates.

The process of selecting a suitable contractor for a building project and tendering processes is covered in section 4.4.

Part of successful planning for a construction project involves efficiently scheduling resources, materials and labour for use at the right time to prevent unnecessary additional costs due to damage, waste, theft, lack of storage and underutilisation.

> **Procurement** Process of agreeing business terms and acquiring goods, products or services from suppliers
>
> **Specifications** Written documents that contain a detailed description of the materials, finishes, workmanship and construction of a building project

Construction scheduling software is often used as part of BIM. (See section 4.7 for further information on Building Information Modelling (BIM).) The following are two commonly used programmes of work:

+ Gantt charts are used to record project start and completion times, and the sequence in which construction activities are scheduled to take place in between.
+ Critical path analysis (CPA) is a decision-making tool used to plan complex building projects. The order and expected duration of activities are plotted using a networking diagram, connected by a series of node points containing critical information.

Software Sequence of digital instructions designed to operate a computer and perform specific tasks

Gantt chart Programme of work used to plan the sequence of building work, delivery of resources and map progress against intended start and completion dates for a construction project

Node points Intersections of lines or pathways in a diagram

Now test yourself   TESTED ◯

8 What is the purpose of TPOs?
9 What project-planning tool uses a networking diagram connected by a series of node points containing critical information?

Revision activity

List nine factors that can influence building design.

Typical mistake

Not all building projects need planning permission; some work can be completed under permitted development rights.

3.5 The concept of the 'whole building', including life cycle assessment

'Whole building' refers to the impact of a building in terms of resources and effect on the natural environment, from the sourcing of materials to manufacturing and construction.

The life cycle of building materials is summarised in Figure 3.3.

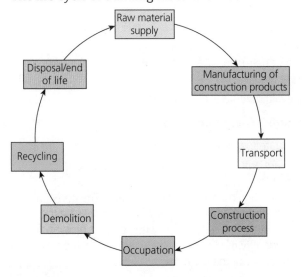

Figure 3.3 The life cycle of building materials

Life cycle assessment

REVISED ◯

The impact that construction materials have on the environment can be calculated using a science-based tool known as 'life cycle assessment' (LCA). This process is summarised in Figure 3.4.

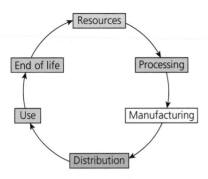

Figure 3.4 Life cycle assessment (LCA)

LCAs:

+ accurately evaluate the effect that materials have at each stage of their life cycle, using data from BIM and other sources of information
+ produce reports that are used in project planning
+ are influenced by environmental regulations and legislation for the whole life cycle of materials from acquisition, use and final disposal to reduce their impact on the environment.

Now test yourself

TESTED ○

10 How can the impact of construction materials on the environment be calculated?

Exam checklist

In this content area, you learned about the following:

+ benefits of good design
+ design principles
+ role of different disciplines involved in design
+ design process from conception to completion
+ the concept of the 'whole building', including life cycle assessment.

Exam-style questions

Short-answer questions

1 What term is used to describe an area of land that has previously been developed or built on, even if there is no physical evidence of earlier use? [1]

2 List the measures used to control and monitor the impact of a building's design, construction and use. [4]

3 Describe the purpose of a waste management plan. [4]

4 Explain the purpose of affordable housing. [3]

5 Determine the impact of over-specifying construction materials and the difficulty of assembling buildings. [5]

6 Describe the advantages of using panelised timber frame as a form of construction. [6]

7 Explain how buildings with architectural or historical interest are protected. [3]

8 Give **four** disadvantages of the brick and block method of construction. [4]

9 Describe the benefits of volumetric construction. [3]

10 Explain the term 'building line' with regards to construction sites. [2]

11 What term is used to describe an area of land that has not been previously developed or built on, above or below ground? [1]

12 Explain the term 'infrastructure' in terms of construction. [4]

13 Explain the process of vernacular construction. [6]

14 Explain the process of starting building work on a domestic building that does not require a full planning application to be submitted to the local planning department. [3]

15 State the role of local authority building control. [1]

16 State **two** responsibilities designers have under the Construction (Design and Management) (CDM) Regulations. [4]

17 Explain the role of a BIM designer. [2]

Extended-response questions

18 Evaluate the benefits of good building design. [12]

19 You are employed as a project manager by an entrepreneur who has recently started a new construction business in the local area. The company has been asked to tender for a project involving the construction of 11 new family homes on a brownfield site. Identify the factors that can impact on the profitability of the project. [12]

20 Explain the process for obtaining planning permission for a two-storey extension on a semi-detached house. Note that the work cannot be completed under permitted development rights. [12]

4 Construction and the built environment industry

4.1 Structure of the construction industry

Construction work is broadly divided into three main categories, as shown in Figure 4.1.

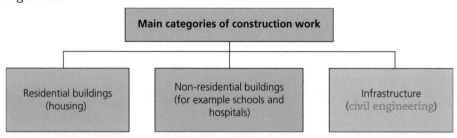

Figure 4.1 The three main categories of construction work

> **Civil engineering** Profession involving the design, construction and maintenance of infrastructure that supports human activities, for example roads, bridges, airports and railways

Business types and roles

REVISED ⬤

The size and scale of a project often determine which type of organisation is involved. For example:
+ established and experienced contractors with proven track records of similar projects are usually on the preferred suppliers list for tenders
+ private clients, who have a smaller amount of money to invest, may appoint subcontractors based on recommendations.

Different business types are involved in the construction sector:
+ Sole traders are individuals who are self-employed.
+ Contractors work for a limited company registered with Companies House. They are responsible for the affairs and day-to-day running of the company but have no personal liability for any financial losses.
+ Subcontractors are self-employed tradespeople who are hired by principal contractors to undertake specific work on building projects.
+ Small, medium and large organisations can be privately owned, a partnership or a corporation.

> **Tenders** Process of inviting bids from contractors to carry out specific projects
>
> **Companies House** Government body that registers and stores information on all the limited companies in the UK and makes it available to the public
>
> **Corporation** Business owned by its shareholders

Client types

REVISED ⬤

Construction work is completed for a variety of different clients:
+ Private clients have construction work carried out for them personally, and not as part of a business.
+ Commercial clients have construction work carried out as part of their business.
+ Public limited companies are managed by directors and owned by shareholders. The business is a separate entity from its owners, so they are protected from any business debt or liabilities.
+ The government is funded by public money to construct schools, universities, hospitals, etc.

> **Commercial** Relating to business; involving buying or selling

The range of work undertaken

REVISED ⬤

The range of work undertaken within the construction industry includes: commercial, residental, industrial, health, retail, recreational and leisure, utilities, transport, new build and retrofit.

The role of building regulations

REVISED

Nearly all new construction work and alterations to existing structures must comply with building regulations. Building regulations are designed to:
+ protect people's health, safety and welfare in and around built environments
+ set industry standards for water and energy use, accessibility and security.

Property owners, for example customers and clients, are ultimately responsible for ensuring that building work complies with relevant building regulations.

Exam tip

A client can be an individual or an organisation. Make sure you can demonstrate your understanding of different types of clients in the construction industry, for example private or government.

Now test yourself

TESTED

1 What is the purpose of building regulations?
2 In which type of business are the owners personally exempt from any liability for financial losses?

Typical mistake

Work in the construction industry is much broader than house building – typically, it can include any of the areas summarised on the previous page.

Revision activity

Create four flashcards with a business type on one side and its definition on the other. With the definitions facing down, choose a card and try to recall the information for each business type.

4.2 How the construction industry serves the economy as a whole

The construction industry's contribution to the UK economy

REVISED

Construction is one of the biggest employment sectors in the UK:
+ It employs around 3.1 million people.
+ Some of the largest contractors generate annual turnovers of several billion pounds.

The construction industry also benefits the UK economy with the activities shown in Figure 4.2.

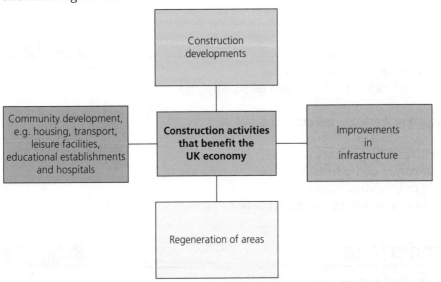

Figure 4.2 Construction activities that benefit the UK economy

Factors that impact growth of the industry

Skilled labour, further and higher education

Market intelligence has identified that the UK construction workforce is an ageing one and that the number of tradespeople retiring from the industry is not being replaced with young people entering the profession with the skills to meet labour forecasts and industry needs.

Where there are skills gaps or shortages of labour due to developments in technology and practice, local and national government often invest heavily in further and higher education to help meet the demand.

Political changes

The extent to which the need for further skilled labour is addressed can be influenced by the government in power at the time and its priorities.

Developments in technology/practice

The UK government is also determined to continue investment in new technologies in the construction sector, to meet the national housing shortage and the need for more affordable homes.

Environmental considerations

Changes are being introduced in the construction industry to improve the environmental climate, including:
+ installing alternative heating systems, for example ground and air source heat pumps
+ installing electric vehicle charging points
+ retrofitting existing buildings with smart technology to improve energy efficiency
+ off-site manufacturing
+ using sustainable materials.

Now test yourself TESTED

3 List the factors that impact the growth of the construction industry.

Revision activity

Make a list of the benefits of the construction industry to the UK economy. Then, cover your list and write down as many of the benefits as you can remember. Finally, check your answers.

4.3 Integration of the supply chain through partnering and collaborative practices

Supply chain

The partners in the supply chain for a building project are shown in Figure 4.3.

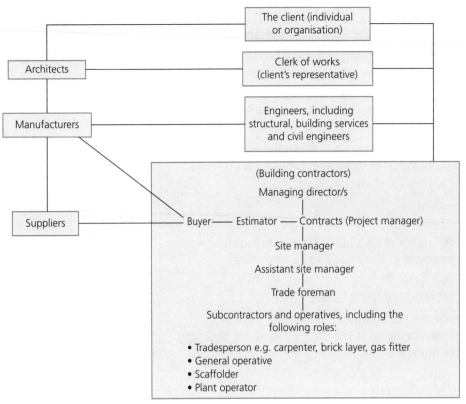

Figure 4.3 Supply chain

Communication through collaborative working and effective planning (inventory management) helps to avoid disruption to the planning/work schedule and increased project costs due to mistakes, increased resources and missed deadlines. Poor planning or communication may also have a negative impact on the reputation of those stakeholders involved.

4.4 Procurement of projects within the construction sector

Procurement

REVISED ⬤

There are different processes that can be adopted by the client to purchase the design and construction of a building. The most common procurement routes are illustrated in Figure 4.4.

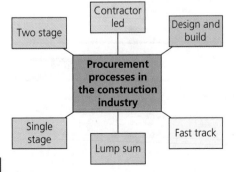

Figure 4.4 Procurement processes in the construction industry

Tendering processes

Tendering is a method used by clients of commercial and industrial projects to find a suitable contractor.

Tender documentation

The client or client's representative is responsible for gathering all the documentation needed for the tender package. The information may include the following documents:
+ letter inviting the contractor to submit a bid for the work
+ outline of the proposal
+ form of tender and timeline to return the completed bid
+ form of contract and conditions, for example terms of payment
+ programme of work.

> **Exam tip**
>
> You should be able to explain the process of tendering in sufficient depth.

Estimation and quotation

An estimate is a prediction of costs for building work, often provided by contractors for domestic clients, based on all the information provided for a job.

A quotation is a fixed price for goods and services offered by contractors or subcontractors to potential clients.

Project cash-flow management

In the terms of contract for a building project, the client and contractor will agree a payment schedule. This will include when the contractor will be paid, for example:
+ up front – an advance payment
+ deposit – a small 'goodwill' payment to help the contractor cover the cost of building materials and labour
+ stage payments – regular payments based on the stage of work completed by the contractor
+ timebound payments (weekly/monthly) – payments made at regular intervals throughout the project, regardless of the progress made with the construction work
+ retention – a sum of money retained by the client for any agreed time; usually a defects liability period between six or 12 months. During the defects liability period the contractor has a responsibility to put right, repair or replace any faults discovered on the project before the retention is paid in full by the client.

Subcontractors will also need to be paid as per their business terms.

If a contractor is not paid regularly and on time, they will often not complete the contract and may take legal action to recover the debt.

> **Typical mistake**
>
> Clients do not award contracts purely based on the lowest tender price. Other factors determining choice may include whether the contractor has completed similar projects and their business reputation.

> **Now test yourself** TESTED
>
> 4 What process is often used to find a suitable contractor for commercial and industrial construction projects?
> 5 Explain the term 'retention' with regards to project cash-flow management.

4.5 Roles and responsibilities of the construction professions and operatives

Construction professions

The roles and responsibilities of construction professions are summarised in Table 4.1.

Table 4.1 Roles and responsibilities of construction professions

Role	Stages they may be involved in	Responsibilities
Architect	Designing and planning	Initially translate information provided by clients and surveys into 2D drawings and 3D digital models, using computer-aided design (CAD)
		Where there is more than one architect, a principal designer (architect) is appointed whose role is to manage, monitor and co-ordinate health and safety during the pre-construction phase
Civil engineer	Designing and planning	Manage the design, construction and maintenance of infrastructure
Ground worker and plant operator	Site preparation and infrastructure	Use construction plant to prepare the site and dig trenches for foundations and to lay services
Building services design engineer	Designing and planning	Consult and advise clients on the concepts and possible approaches for supplying the following services: + heating + ventilation + air conditioning + renewable energy + sustainable technologies + lighting + fire and security systems
Building services engineer technician	Designing and planning	Assist the project team with design solutions, specifications and planning for building services engineering systems
	First and second fix	Supervise specialist contractors
		Take the lead on health and safety
		Commission systems
		Monitor quality control, record progress and report to the project team during installations
Building services engineering site manager	First and second fix	Oversee the installation of complex environmental systems, such as heating, lighting and electrical power
Facilities manager	Operational	Take responsibility for the operation, servicing and maintenance of building services once the building work has been completed, signed off and handed over to the client
Client representative (clerk of works, CoW)	All stages of the construction phase	Usually appointed by the principal designer or the client to represent the client onsite during the construction phase
		Work closely with construction staff, surveyors and engineers to make sure plans and specifications are followed properly
Contract manager	Planning and all stages of construction	Employed by contractors to assist in preparing tenders for clients and securing future business

Check your understanding and progress at **www.hoddereducation.co.uk/myrevisionnotes**

Now test yourself

TESTED

6 Who is responsible for checking drawings and specifications to ensure work is completed to the standards agreed in the building contract and reporting their findings to the client?

Construction operatives

REVISED

The roles and responsibilities of construction operatives are summarised in Table 4.2.

Table 4.2 Roles and responsibilities of construction operatives

Role	Stages they may be involved in	Responsibilities
Joiner	Off-site manufacturing during the construction phase	Based in a workshop, manufacture purpose-made items to order, e.g. doors, windows, frames and staircases
Carpenter	First and second fix	Usually work on construction sites at the first-fix stage installing floors, walls, roofs and stairs, and at the second-fix stage fitting skirting, architraves, doors and kitchens
Plasterer	Internal finishes – between first and second fix External finishes – after first fix	Apply smooth and textured finishes to internal and external walls using a range of materials, such as gypsum, cement or lime, and modern materials such as polymer or acrylic renders
Bricklayer	Structural	Set out, build and repair walls, piers and archways for domestic and commercial projects
Plumber	First and second fix	Install, maintain and repair water, heating and drainage systems in new and existing buildings
Decorator	Finishing	Prepare surfaces such as walls, ceilings, metal and woodwork for decorative and protective finishes
Non-skilled operative	Construction phase	Undertake manual labouring tasks to support tradespeople and other skilled workers on site

Now test yourself

7 What type of work does a civil engineer carry out?

TESTED

4.6 The role of continuing professional development (CPD) in developing the knowledge and skills of those working in the sector

Some professionals and operatives have legal responsibilities to belong to professional, accredited or certified organisations in order to actively continue with their job roles. This often involves maintaining professional standards and keeping records of continuing professional development (CPD).

Types of CPD include: formal, informal, qualifications, work experience, self-learning and chartered.

> **Accredited** Officially recognised as meeting professional quality standards
>
> **Continuing professional development (CPD)** Process of maintaining, improving and developing knowledge and skills related to one's profession in order to demonstrate competence

Importance of CPD

REVISED

CPD helps to maintain occupational competence and best practice. There are many benefits for employers and individuals in upskilling, including:
+ protecting clients, customers and the public from construction activities
+ keeping up to date with changes in regulations, product development and technological advancements
+ developing product knowledge
+ working more efficiently
+ improving knowledge and skills
+ enhancing the company image
+ career progression.

Workforce planning

REVISED

Workforce planning is the process a business may follow to assess the current and future needs of its employees against the organisation's demands. CPD may focus on a particular area of weakness that has been identified, such as:
+ legislation
+ management and supervision
+ health, safety and welfare
+ digital technology
+ conservation and refurbishment
+ sustainability
+ maintenance
+ tools and equipment
+ industry standards and best practice.

Providers of CPD

REVISED

Providers of CPD include: professional bodies, e.g. Chartered Institute of Building (CIOB); accreditation bodies, e.g. British Standards Institution (BSI); certification bodies, e.g. Gas Safe; manufacturers, e.g. Festool; and in-house toolbox talks, e.g. ladder safety.

4.7 Building Information Modelling (BIM)

Building Information Modelling (BIM) uses smart technology to allow effective and efficient collaboration between designers and the construction team at every stage of a building project.

Aspects of BIM and the effect it has on real-time project delivery

REVISED ◯

Designers usually start the BIM process by translating information captured from a construction site and the client's drawings into digital 3D models of the building and the infrastructure around it.

BIM:
+ illustrates every detail of a project in graphical form (drawings) and non-graphical form (written information)
+ records the relationships between components and how they all fit together
+ enables information to be shared easily with all members of the project team.

BIM government levels 0–3

REVISED ◯

To bring clarity to how BIM is used in a project, the government has defined a number of levels, as described in Table 4.3.

Table 4.3 BIM levels

BIM level	Description
Level 0	Designers use computer-aided design (CAD) to produce drawings and plans, but there is no digital collaboration between stakeholders.
Level 1	A Common Data Environment (CDE) is established, allowing information to be stored centrally and accessed by the whole project team.
Level 2	Any CAD software can be used at this level, but it must be capable of being exported to common file formats.
	The project team will be able to work collaboratively using different systems throughout a construction project. This is the minimum level required by government for public construction projects.
Level 3	This allows an 'open data' standard, so that all stakeholders are able to work simultaneously on the same project, from anywhere in the world.

The collaborative role of BIM

Digital Plan of Work (DPoW)

At the design stage, BIM is used to generate documentation for construction specifications and schedules for the building phase of the project, referred to as the Digital Plan of Work (DPoW).

Employer's Information Requirements (EIR)

At the start of a project, the client usually meets with the designer and contractor to discuss the information they want to receive at each stage up until handover, and how this will be shared with them through BIM. This detail is recorded in a document known as the Employer's Information Requirements (EIR).

Common Data Environment (CDE)

This allows information to be stored centrally and accessed by the whole project team.

> **Exam tip**
>
> It is likely that there will be questions about BIM in the exam. Make sure you can describe the benefits of the system and how it is used to collaborate on construction projects.

> **Now test yourself**
>
> 9 At which government BIM level can any CAD software be used to export common file formats?
>
> TESTED

> **Typical mistake**
>
> BIM is not just used to create construction drawings and 3D models. It is also used to collaborate on projects using other types of construction documentation, for example specifications and schedules.

4.8 PESTLE factors

PESTLE is an acronym used to describe factors that are beyond the control of a business, as outlined in Table 4.4.

Table 4.4 PESTLE factors

Political	Political situations could affect local and national government spending in the construction sector.
Economic	This includes higher taxation, an increase in interest rates or a slowdown in the economy due to people not spending money.
Social	Population demographics and the movement of social and community groups can influence the type of buildings and structures constructed in particular areas to meet the needs of the community.
Technological	Technological innovations could affect how a building or structure is designed and constructed, e.g. augmented reality.
Legal	Laws made by Parliament can influence the planning, design and construction of buildings, e.g. changes in legislation following the Grenfell Tower fire in 2017.
Environmental	Companies have corporate social responsibility (CSR) and must protect the environment by reducing the carbon emissions produced during manufacturing and construction.

> **Exam tip**
>
> Revisit section 3.1 to recap corporate social responsibility (CSR) and help build your answers for questions about PESTLE factors.

> **Typical mistake**
>
> PESTLE factors *cannot* be controlled by a business.

> **Now test yourself**
>
> TESTED
>
> 10 What does the letter L stand for in the acronym PESTLE?

> **Revision activity**
>
> Practise writing down what each letter stands for in the acronym PESTLE and provide two current examples of each factor.

Exam checklist

In this content area, you learned about the following:

+ structure of the construction industry
+ how the construction industry serves the economy as a whole
+ integration of the supply chain through partnering and collaborative practices
+ procurement of projects within the construction sector
+ roles and responsibilities of the construction professions and operatives
+ the role of continuing professional development (CPD) in developing the knowledge and skills of those working in the sector
+ Building Information Modelling (BIM)
+ PESTLE factors.

Exam-style questions

Short-answer questions

1 Describe the process of open tendering for the procurement of projects within the construction sector. [3]

2 Identify and explain the condition that is agreed in the terms of a building contract between a client and a contractor that imposes fines on the contractor if work is not finished by the completion date. [2]

3 Explain how appointing a clerk of works will benefit a client in the successful completion of a construction project. [3]

4 List the benefits of CPD. [4]

5 Describe the process of identifying corrective work needed to building work at the end of a project, before handing over to the client. [2]

6 Explain the term 'retention' with regards to a building contract. [2]

7 State the **three** categories into which construction work is broadly divided in the UK. [3]

8 PESTLE factors are things that are beyond the control of a business. What does the acronym PESTLE mean? [3]

9 Describe the responsibilities of a facilities manager. [2]

10 Explain the term 'renewable energy'. [1]

11 Explain the benefits of Building Information Modelling (BIM). [2]

12 A construction project is valued at £138,000. The building contract stipulates a penalty clause of 5 per cent for every day that the project runs over. The construction project takes a week and three days longer than expected. How much money will the client withhold from the contractor? [2]

13 List **four** different types of tendering process used in the construction industry. [2]

14 State **one** advantage of being a limited company. [1]

Extended-response questions

15 Explain how the construction industry serves the UK economy. [9]

16 The UK government passed the Climate Change Act in 2008, with a target to reduce carbon emissions to net zero (carbon neutral) by 2050. Describe the changes being introduced in the construction industry to improve efficiencies and address the need for a cleaner environment. [9]

5 Construction sustainability principles

5.1 Sustainability when planning and delivering a construction project

The importance of sustainability

Sustainable design and construction take account of:
+ the resources used in construction
+ the environmental, social and economic impacts of the construction process itself
+ how buildings are designed and used.

When planning and delivering a construction project, sustainability is achieved by:
+ using renewable and recyclable resources
+ sourcing materials locally
+ protecting resources
+ reusing and refurbishing materials
+ reducing energy consumption and waste
+ creating a healthy and eco-friendly environment
+ protecting the natural environment.

Two key aspects when considering sustainability in construction are:
+ life cycle assessment – assessing the total environmental impact of a building, considering all stages of the life of the products and processes used in it
+ carbon footprint – the amount of carbon dioxide released into the atmosphere because of construction activities.

Global construction accounts for 38 per cent of total global carbon emissions:
+ Half of all carbon emissions are embodied in buildings, meaning they are caused by the manufacturing of materials and the construction process.
+ Local sourcing of materials, resource reuse and refurbishment of materials can all help in the reduction of global emissions and reduce a building's carbon footprint. This includes using recycled bricks, tiles/slates and timber products.

Sustainability assessment methods

Various assessment methods are used to determine how well a building performs against environmental, social and economic standards. These include:
+ Building Research Establishment Environmental Assessment Method (BREEAM), which assesses the environmental performance of buildings.
+ Leadership in Energy and Environmental Design (LEED), which rates the design, construction, operation and maintenance of green buildings.
+ Timber Research and Development Association (TRADA), which looks at the use of wood in the built environment.
+ WELL Building Standard, which considers those aspects of the built environment that affect human health and wellbeing.

Check your understanding and progress at **www.hoddereducation.co.uk/myrevisionnotes**

The following organisations can advise on sustainability matters:
+ Energy Saving Trust: an independent organisation – working to address the climate emergency.
+ Timber Research and Development Association (TRADA).

5.2 Types of sustainable solutions

Sustainability involves a commitment to environmental, economic and social objectives, as described in Figure 5.1.

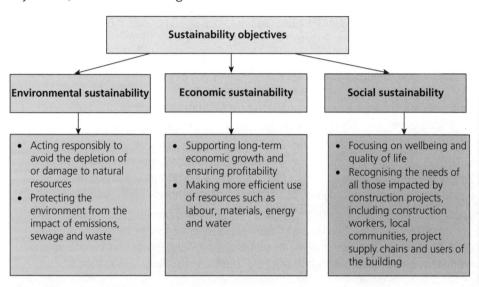

Figure 5.1 Environmental, economic and social sustainability objectives

Different types of sustainable solution are described in Table 5.1 (see also section 5.8 on recycling).

Green roof Sustainable roof system that involves installing additional waterproof membranes and drainage mediums, onto which soil is added to allow growth of vegetation; this protects and insulates the building and reduces its environmental impact

Water undertaker A water company that has the statutory duty to supply water/and or sewage services within a geographical area

Table 5.1 Sustainable solutions

Sustainable solution	Description
Prefabricated construction	Prefabricated construction combines pre-engineered units to form major elements of a building. They are manufactured in factories and then transported to site where they are assembled.
Self-healing concrete	Self-healing concrete contains the spores of limestone-producing bacteria and a food source. When cracks occur, moisture in the air causes the spores to germinate. The reactivated bacteria then eat the food source and excrete calcite to heal the crack.
Green roofs	A green roof, also known as a living roof, is an attractive and sustainable roof system that involves installing additional waterproof membranes and drainage mediums, onto which soil is added to allow growth of vegetation.
Smart glass	Smart glass (also known as switchable glass) changes from transparent to translucent (and vice versa) when exposed to specific levels of voltage, heat or light.
Electrochromic glass	Electrochromic glass is an electrical glass product that allows you to control the amount of light and solar radiation that can enter a space through the glazing.
Soakaway	A soakaway is a hole dug into the ground that is filled with coarse stone and rubble or plastic crates. It allows water to filter through it and soak into the ground. The use of a soakaway reduces the need for water to be processed by the water undertaker.
Reedbed	A reedbed is constructed wetland that uses natural filtration and biological processes to break down organic matter in waste water and sewage effluent.

Now test yourself

TESTED

1 List **four** sustainable construction solutions.

73

5.3 Environmental legislation

The purpose of environmental legislation is to protect, preserve the environment and control hazards to health.

Key environmental legislation is summarised in Table 5.2.

Table 5.2 Environmental legislation

Legislation	Key aims/requirements
Environmental Protection Act 1990	+ Defines legal responsibilities for the management of waste and pollution + Places a duty on local authorities for collecting waste + Lays out the duty for businesses to handle waste safely
Climate Change Act 2008	Set a target for the UK to reduce its greenhouse gas emissions by 80 per cent by 2050, compared to 1990 levels (target was updated in 2019 to 100 per cent – net zero)
Clean Air Act 1993	+ Covers a range of topics, including prohibition of dark smoke from chimneys and the requirement for new furnaces to be smokeless + Allows local authorities to declare the whole or part of the district of the authority to be a smoke control area, where it is an offence to emit smoke from the chimney of any building, furnace or fixed boiler
Water Act 2014	Aimed to: + reform the water industry to make it more innovative and responsive to customers + increase the resilience of water supplies to natural hazards such as droughts and floods + address the availability and affordability of insurance for households at high risk of flooding
Building Regulations 2010	Set the standards for the design and construction of new buildings and alterations to buildings (guidance on compliance is published by the Ministry of Housing, Communities and Local Government in the form of approved documents that provide general advice on the performance expected of materials and building work and practical solutions to some common building situations – see also section 7.3)
Control of Substances Hazardous to Health (COSHH) Regulations 2002	+ Aim to protect people from ill health caused by exposure to hazardous substances (manufacturers and suppliers of hazardous substances produce safety data sheets that contain important information about how products should be transported, used, stored and safely disposed of after use) + Provide information on measures that must be taken to ensure substances do not pose an environmental hazard, e.g. chemicals being discharged or leaked into water sources
Waste Electrical and Electronic Equipment (WEEE) Regulations 2013	+ Implement the provisions of the EU WEEE Directive in UK law + Aim to support sustainable production and consumption through the collection, reuse, recycling, recovery and treatment of end-of-life electrical and electronic equipment (EEE)
Hazardous Waste (England and Wales) Regulations 2005	Restrict the production, movement, receiving and disposal of hazardous waste, such as fluorescent tubes, refrigerators and asbestos
Control of Pollution (Oil Storage) (England) Regulations 2001	+ Designed to reduce incidents of oil escaping into the environment + Require anyone in England who stores more than 200 litres of oil to provide a more secure containment facility for tanks, drums, bulk containers and **mobile bowsers**
Site Waste Management Plans Regulations 2008	Specify how a building site is required to manage its waste

> **Mobile bowsers** Wheeled trailers fitted with a tank for carrying oil

Typical mistake

Students often miss key points when responding to a question. Make sure you reread the question after completing your initial response, to ensure all aspects have been addressed. For example, if you are asked to detail an item of environmental legislation ensure you include the full title, date and an overview.

Revision activity

Create a list of environmental legislation and describe the purpose of each piece.

Check your understanding and progress at **www.hoddereducation.co.uk/myrevisionnotes**

5.4 Environmental policies and initiatives

Conservation of fuel and power Approved Document L

This document provides guidance on the energy efficiency of different structures. It is updated regularly to reflect developments in building materials and new technologies, therefore people must ensure they are using the most up-to-date version. Approved Document L is divided into four parts, with each providing specific details on a particular building type:

+ L1A – new dwellings
+ L1B – existing dwellings
+ L2A – new buildings other than dwellings
+ L2B – existing buildings other than dwellings.

Hazardous Waste Regulations

The Hazardous Waste Regulations are a set of rules that control and track the movement of hazardous waste in England and Wales. They came into force in July 2005 and implement the Hazardous Waste Directive. The regulations require organisations that produce or handle hazardous waste to register with the Environment Agency, classify the type of waste, store it safely, and dispose of it through an authorised waste carrier. The regulations aim to reduce the risks and harm caused by hazardous waste. Section 34 of the Environmental Protection Act 1990 states there is a duty of care to ensure that waste is managed and disposed of properly:

+ If a business transports waste, either for itself or for someone else, it needs to register with the Environment Agency as a waste carrier.
+ If a business's waste is being collected by someone else, the business must ensure that the carrier is registered.

There are two types of licence, depending on the type of waste to be transported:

+ An upper-tier waste carrier transports other people's waste on a professional basis.
+ A lower-tier waste carrier was either previously exempt from registration or carries its own (non-construction/non-demolition) waste on a regular and normal basis.

Home Quality Mark

The Home Quality Mark is a voluntary assessment and certification scheme. It was developed by the Building Research Establishment (BRE) and it is used to assess the impacts of new-build homes. The mark provides a rating out of five stars and other indicators which are assessed against particular criteria, including the home's surroundings, the comfort of the home and other considerations such as construction energy use, water use, site waste and considerate construction methods.

Key requirements of environmental regulations

REVISED

Energy Performance of Buildings Directive

The Energy Performance of Buildings Directive is an EU legislative instrument that aims to reduce the carbon emissions produced by buildings. It requires the:

+ production of an energy performance certificate (EPC) whenever a building is sold, rented out or constructed
+ production of an energy certificate for large public buildings, which must be displayed in a prominent place
+ regular inspection of air-conditioning systems and boilers.

Revision activity

Visit the following website and produce a document detailing each aim of the quality mark:

www.ukbuildingcompliance.co.uk/wp-content/uploads/2018/02/home-quailty-pdf.pdf

5.5 Environmental performance measures

Environmental performance measures are summarised in Table 5.3.

Table 5.3 Environmental performance measures

Environmental performance measure	Details
Source and use of materials	Sourcing materials locally has a lower environmental impact as it uses less transportation and therefore causes fewer carbon emissions. This will reduce the carbon footprint of the project.
Energy source and consumption	The energy efficiency of a building depends on its use, design, orientation, location and the materials used in its construction. For example: + a south-facing building will be more efficient due to solar gains + materials with a lower U-value will lose less heat.
Water source and consumption	To improve a building's environmental impact, it is important to minimise water usage. This can be achieved, e.g. by installing flow-reducing valves, low-flush WCs and infrared controls.
Pollution and waste processing	Construction that generates large amounts of waste or causes pollution to land, air or water will have a negative environmental impact. During construction, it is important that this is considered and offset through material recycling, a site waste management plan and putting controls in place in case of an environmental accident.
Transport	Transporting construction materials over long distances will have a negative effect on the environment due to vehicle emissions.
Destruction and disposal	Well-planned management of demolition means that building materials used during construction can be recycled, thereby reducing environmental impact.
Radioactive waste	Radioactive waste includes any material that is either intrinsically radioactive, or has been contaminated by radioactivity. In the UK, radioactive wastes are classified according to the type and quantity of radioactivity they contain and how much heat is produced.
Flexibility	Designing flexible buildings for easy change and adaptation will assist the main concept of sustainability by reducing material and energy consumption as well as environmental pollution.
Durability	Designing and constructing buildings with the ability to withstand wear, pressure and damage is key to sustainability.
Sustainable landscaping	Sustainable landscaping is the implementation of landscape strategy to offset environmental impacts of man-made industry and development while recognising the value of ecosystems.

Schemes used to certify environmental performance

REVISED

Several schemes can be used to certify levels of environmental performance in construction, for example:
+ BREEAM and LEED (see section 5.1)
+ Passivhaus ('passive house' in English) – an energy performance standard intended primarily for new buildings, which ensures that buildings are so well constructed, insulated and ventilated that they require little energy for heating or cooling.

Now test yourself

2　List **three** things that should be considered when choosing construction materials.

TESTED

5.6 Principles of heritage and conservation

Listed buildings

REVISED

Listed building consent is required for all works of demolition, alteration or extension to a listed building that affect its character. This is to ensure that the impact of any proposed changes is reviewed before they are approved. This is usually carried out by a conservation officer within the local authority planning department.

Buildings are listed in their entirety, even though some parts may be more important than others. The designation regime is set out in the Planning (Listed Buildings and Conservation Areas) Act 1990.

There are three types of listed status for buildings in England and Wales:
+ Grade I – buildings of exceptional interest
+ Grade II* – particularly important buildings of more than special interest
+ Grade II – buildings of special interest, warranting every effort to preserve them.

The Heritage Protection Bill is a legislative and policy framework that protects the historic environment. It requires consents and permissions to protect England's heritage via a balanced, democratic and informed approach to managing changes in historic places.

When carrying out work on listed buildings, it is important that you do not use modern repair methods on traditional construction. For example:
+ Using cement on older buildings made from materials such as lime mortar can cause damage.
+ Traditional cast-iron soil stacks should be replaced like for like, and not substituted with modern plastic pipework.
+ Any original architectural features such as doors, decorative stonework, fireplaces or windows should not be altered.

> **Typical mistake**
>
> Students sometimes do not contextualise their answer in relation to the question, so check you have done this. This can be achieved by including the key words and information from the question.

Conservation areas

REVISED

Local planning authorities have the power to designate any area of special architectural or historic interest as a conservation area, where the character or appearance should be preserved. The special character of these areas is not just made up of buildings but can also be defined by features that contribute to particular scenic views, such as woodland or open spaces.

Town and Country Planning Act 1990

REVISED

Under the Town and Country Planning Act 1990, planning permission needs to be sought when carrying out work to:
+ build a new property
+ increase the size of an existing property
+ make significant alterations to an existing property
+ change the use of an existing property.

> **Now test yourself**
> TESTED
>
> 3 Explain the **three** types of listed building.

5.7 Lean construction

Principles of lean construction

REVISED

Lean construction is a construction methodology that aims to minimise waste in terms of costs, materials, time and effort, while maximising productivity and value. The principles of lean construction are summarised in Figure 5.2.

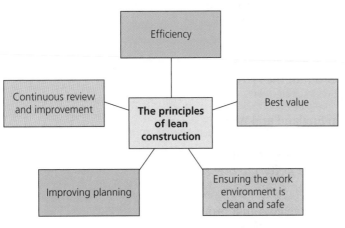

Figure 5.2 The principles of lean construction

Techniques to maximise value and minimise waste

REVISED

A common example of a lean construction technique is just-in-time deliveries. This is a method of providing the required materials for a project in precisely the correct order and quantity at exactly the right time for installation.

Another lean construction technique is artificial intelligence (AI), which is commonly used in the construction sector to develop safety systems for worksites.

Other techniques that can be used to maximise value and minimise waste are:
+ reducing errors during the construction stage
+ accurately measuring materials and efficient ordering
+ recycling and repurposing materials throughout the duration of the project.

Advanced manufacturing techniques

REVISED

The manufacturing industry is constantly evolving. Increased use of digitalisation has resulted in processes that are more efficient, effective and responsive and which rely less on human effort:
+ Computer numerical control (CNC) is used in manufacturing as a method for controlling machine tools using software. It allows data produced in CAD programs to control automated operations, such as milling, lathing, routing and grinding. This results in rapid, accurate and repeatable machining of bespoke components.
+ Off-site construction describes the planning, design or assembling of building parts in a factory. These are then transported to site where final assembly is completed.

> **CAD programs** Computer-aided design software used to produce design and technical documentation

Check your understanding and progress at **www.hoddereducation.co.uk/myrevisionnotes**

5.8 Waste management

The waste hierarchy

The waste hierarchy sets out the order in which actions should be taken to manage waste, from the most to least preferable in terms of environmental impact. There are five actions, as shown in Table 5.4.

> **Anaerobic digestion**
> Process by which bacteria break down organic matter

Table 5.4 The waste hierarchy

Step	Aim	How this is achieved
1 Reduce	Produce less waste	+ Use fewer materials during design and manufacture. + Keep products for longer. + Choose products with less packaging. + Use fewer hazardous materials.
2 Reuse	Extend the lifetime during which products are used	+ Use a product multiple times, rather than single use. + Repurpose products or their parts at the end of their life cycle by checking, cleaning, repairing or refurbishing them.
3 Recycle	Make new products from those no longer used	+ Turn waste into new substances or products. + Compost the material (provided it meets quality protocols).
4 Recover	Turn products no longer used into a source of energy	Use waste products as fuel to provide heat and power; other methods include **anaerobic digestion**.
5 Landfill	Dispose of and destroy products	Dispose of and incinerate products without energy recovery.

Circular economy principles

The concept of a circular economy aims to maximise total material resource efficiency by:
+ designing out waste
+ designing for adaptability
+ building in layers
+ selecting materials
+ designing for deconstruction.

Site waste management plans

While no longer a legal requirement, a site waste management plan (SWMP) is an important document for setting out how waste will be managed and disposed of during a construction project.

It should be compiled by the design team, contractor and subcontractors and refer to the waste hierarchy to reduce the volume of waste to landfill and increase the use of sustainable materials. This ensures compliance with environmental legislation, improves resource efficiency and increases profitability.

Reducing use of pollutants in construction projects

Under the Environmental Damage (Prevention and Remediation) (England) Regulations 2015, businesses are made financially liable for any damage they cause to land, air, water and biodiversity in England. To ensure compliance, a business should:
+ never burn waste materials
+ adopt hybrid technology, such as diggers/excavators
+ use water sprays or sprinklers to control some types of dust

+ keep materials secure
+ keep the site clean
+ cover up drains.

Waste segregation

Waste segregation means dividing waste into different categories for efficient disposal. Not only does this support the waste hierarchy and protect the environment, by offering opportunities to reuse and recycle waste before resorting to landfill, but it also offers cost savings to construction businesses. Sorted waste is cheaper to dispose of, and some types of waste can even be sold as a source of income.

Waste on a construction site is usually segregated into the following streams:
+ general (for example insulating materials that do not contain asbestos)
+ hazardous (for example asbestos)
+ clean fill (material that can be recycled or reused in future construction projects)
+ hard fill (for example soil, concrete, bricks and blocks)
+ plastic
+ metal
+ wood.

Recycling

Different materials can be recycled in various ways. Scrap or waste metal, waste plastic and offcuts of timber can be reprocessed into many different products. Undamaged bricks and blocks can be reused, and waste from cutting operations can be crushed and used as aggregate.

> **Now test yourself**
>
> 4 Explain why hazardous waste should be segregated.
> 5 List **three** recycled materials.

5.9 Energy production and energy use

Types of energy production

Section 2.3 provides an overview of energy production, covering both renewable and non-renewable energy sources.

Wind energy

The natural kinetic energy of the wind can be used to drive an aerodynamic bladed turbine, rotate a generator and produce electricity.

Often, many wind turbines are grouped together to form a wind farm and provide bulk power to the National Grid.

Turbines can be installed on land or offshore where there is a reliable source of wind.

Solar (photovoltaic) energy

Solar energy utilises photovoltaic (PV) panels to convert the sun's radiation into electricity. The panels can be installed on the roofs of buildings or in larger-scale solar farms. Solar farms can deliver bulk power to the National Grid.

To maximise the energy generated by PV panels, is it important to install them carefully, for example where there is enough sunlight and facing an appropriate direction.

Nuclear power

Nuclear power plants use reactors to split atoms, causing a large amount of thermal energy to be released. This heat is then used to heat water to produce steam. The steam turns a turbine and generator to produce electricity.

Hydroelectric energy

Hydroelectric power plants produce energy by channelling running water through a turbine connected to an electrical generator. The electricity is then fed into the National Grid for distribution.

Hydroelectric power plants can utilise:
+ the natural flow of a river as it falls from a greater to a lesser height
+ artificial reservoirs and dams, which hold back water and release it as required.

Wave and tidal energy

Wave and tidal generators are similar to wind generators, except that they use the ocean's current to rotate the turbine instead of wind.

The amount of energy created is determined by the wave's height, speed, wavelength and water density.

Wave power is much more predictable than wind power, and it increases during the winter, when electricity demand is at its highest. Tidal energy is also predictable and consistent.

Hydrogen

Hydrogen can be used in fuel cells to generate power. It:
+ is a clean fuel that, when consumed in a fuel cell, produces only water
+ can be produced from a variety of domestic resources, such as natural gas, nuclear power, biomass and renewable power like solar and wind
+ is zero-carbon, provided it is created in a process that does not involve the burning of fossil fuels.

Biomethane

Biomethane is a naturally occurring gas produced by organic matter such as dead animal and plant material. It is chemically identical to natural gas, so can be used for the same applications, as well as to fuel vehicles.

Biomass

Biomass is organic, meaning it is made of material that comes from living organisms such as plants and animals. It can be used in power stations and for domestic biomass boilers.

Carbon capture and storage (CCS)

Carbon capture and storage (CCS) processes remove carbon dioxide (CO_2) that would otherwise be emitted from fossil-fuel power stations and industrial processes.

> **Exam tip**
>
> When a question asks you to explain the working principles of a technology, it is important that you provide a full and detailed explanation.

> **Hydroelectric power**
> Form of renewable energy that uses the power of moving water to generate electricity

5 Construction sustainability principles

Exam tip

The questions in the exam are arranged in order of gradually increasing difficulty.

Exam-style questions

Short-answer questions

1 Describe the purpose of sustainability assessment methods and list **three** types. [4]

2 Explain the purpose of a reedbed. [2]

3 Describe the purpose of the Control of Pollution (Oil Storage) (England) Regulations 2001. [1]

4 Describe the enforcement of building regulations. [2]

5 Explain **two** examples of how construction materials can be reused following the demolition of a building. [2]

6 Describe listed building consent and where this is required. [3]

7 Describe the benefits of just-in-time deliveries. [4]

8 List **three** new resources that can be produced from recycled timber. [3]

9 Describe what is meant by the term 'fossil fuel'. [4]

10 Explain the term 'reduce' in the waste hierarchy. [5]

11 List **three** zero-carbon technologies. [3]

12 Explain the approximate heat loss through parts of a typical detached house and how this can be avoided. [5]

13 Describe how a nuclear power plant works. [4]

14 List the waste hierarchy in the correct order. [5]

Extended-response questions

15 Explain what biomass is and how it is sustainable. [9]

16 Describe the purpose and key components of a solar thermal hot-water system. [9]

17 A client wants to incorporate renewable energy systems into an existing two-storey office. The building has an existing plant room and a south-facing pitched roof. The building is close to neighbouring properties with little outdoor space. Evaluate the different types of renewable energy systems and suggest the most suitable for this project. [12]

6 Construction measurement principles

6.1 Accurate and appropriate measurement

Construction measurement principles are tried and tested methods and rules, which if applied correctly will ensure accuracy in:
+ calculating quantities for a project
+ setting out and building a structure.

Methods of measurement vary depending on the work task and the operational phase of the construction process. For any size of construction project, accurate and appropriate measurements are used to follow the steps shown in Figure 6.1.

Exam tip

Questions about measurements often include a scenario. Spend some time imagining yourself in the scene before answering the question.

Figure 6.1 Steps in a construction project

The benefits of accurate measurement
REVISED

Benefits to the client

A client expects that:
+ work should be completed within an agreed budget and timescale
+ the design brief should be wholly fulfilled.

Design brief Document for the design of a project developed in consultation with the client/customer

Benefits to project success

Use of accurate and appropriate measurement methods is vital if a building is to be constructed exactly as shown on the drawings and in complete accordance with specifications.

Benefits to the contractor

Accurate measurement allows for accurate costing of a project. This can mean:
+ the contractor can manage profitability and cash flow more successfully to maintain consistent business activity and company growth
+ surplus and waste are reduced so that profit margins are maintained.

Revision activity

Write a short report of around 100 words describing how accurate measurement affects the planning, design and construction phases of a building.

Choosing appropriate methods of measurement
REVISED

Throughout the stages of a construction project, different work activities require different methods of measurement. Table 6.1 gives some examples of appropriate methods of measurement and the benefits of applying them accurately.

Table 6.1 Appropriate methods of measurement and the benefits of applying them accurately

Construction activity	Measurement method	How the method is used	Benefits of accuracy
Site survey	Calculation of area	A building must: + fit on a proposed site + allow for paths, roadways, parking and landscaped areas + ensure adequate space between the new building and existing structures.	+ Confirms that the proposed project is feasible + Allows for an appropriate design concept to be created
Locating services on a site	**Linear measurement**	The position of existing services must be established. Records regarding the position and route of services are used to accurately measure from stated positions on site.	+ Allows site work to proceed safely + Reduces the risk of accidental damage to existing services, so avoids causing delays and adding costs
Groundworks including removal of topsoil and excavations	Calculation of volume	The volume of soil to be excavated, e.g. for foundations or installation of drainage, is measured in cubic metres (m³).	+ Establishes the type and number of machines needed for excavations + Establishes the timescale of the operation
Correctly locating the new structure on the plot	Linear measurement	Accurate measurement from given reference points is used to position the building in the correct location according to working drawings.	+ Ensures the structure will not interfere with other elements of the development or contravene planning requirements + Ensures correct location – a wrongly located building will need to be demolished and rebuilt in the correct position, at great expense
Building the new structure	Linear measurement	Accurate linear measurement is needed to: + set out the outline of a building + set out the position of internal walls, doors and windows + confirm vertical dimensions to establish floor and roof heights.	Ensures the building is constructed exactly as designed
	Calculation of area	This confirms quantities of materials required for plastering, bricklaying, roofing, etc.	+ Avoids shortfalls of materials leading to delays + Avoids excess materials leading to waste
	Calculation of volume	This identifies requirements in cubic metres for materials such as concrete or **screed** required for solid floors.	

Linear measurement Distance between two given points along a line

Screed Levelled layer of material (often sand and cement) applied to a floor or other surface

Exam tip

Take care not to confuse types of measurements with the units in which they are measured. For example, 'area', 'volume' and 'linear' are *types* of measurement; the units they are expressed in are m², m³ and mm respectively.

Revision activity

Select two construction activities from Table 6.1. For each, explain why accurate measurement is important.

Check your understanding and progress at **www.hoddereducation.co.uk/myrevisionnotes**

Costing techniques for construction projects

Job costing

Job costing is usually applied to specific client requirements for a distinct project or job. It works by:
+ analysing the job in detail
+ breaking down costs for each job element
+ tracking and recording those elements as the job proceeds.

Elements are broken down into groups consisting of:
+ labour
+ materials
+ tools and equipment
+ overheads.

> **Overheads** Costs of running a business that are not directly related to production

Typical mistake

Job costing is sometimes confused with process costing, where the steps to complete a job are identified and an average cost is applied to each step based on past experience.

Job costing can sometimes include a 'cost reimbursement' condition, where the client agrees to meet the cost of job elements that cannot be accurately calculated at the start of a project, for example if the design choice of certain items such as windows has not been finalised.

Batch costing

This can be viewed as an extension of job costing, since a batch (like a job) is broken down into component materials to be costed individually.

Activity costing

Activity costing assigns costs to specific activities within a project rather than the whole job or process. Identifying the many specific activities required to produce a building provides costing data that can feed into the overall costing process.

For a specific activity, the associated costs for materials, labour, equipment and appropriate overheads can be used to:
+ cost the current project
+ inform the costing process for future projects with similar characteristics.

Life-cycle cost analysis

Life-cycle cost analysis is used to estimate the overall costs of a building throughout its entire life cycle. This method supports:
+ analysis and comparison of alternative building concepts during the design stage
+ evaluation of which design is most economically and environmentally beneficial over time
+ analysis of everything required throughout the life cycle of the building.

Life-cycle cost analysis of a building may require consideration of the aspects shown in Figure 6.2.

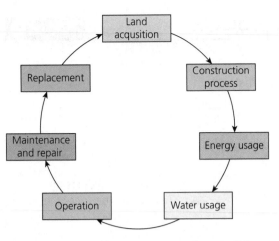

Figure 6.2 Life-cycle cost analysis for a building

Now test yourself

TESTED ⬤

1 In what ways does a client benefit from accurate measurement?

2 Identify the benefits of accurate measurement to a contractor.

3 Define cost reimbursement.

4 Which costing technique would be most appropriate when retrofitting a ground source heat pump? Explain your selection.

6.2 Standard units of measurement and measurement techniques

Units of measurement

REVISED ⬤

Metric units

The UK construction industry uses metric units as standard.

The standard metric units of measurement and their abbreviations in ascending order of size or value are shown in Table 6.2.

Table 6.2 Metric units of measurement

Measurement	Metric units (and abbreviations)
Length	millimetres (mm)
	centimetres (cm)
	metres (m)
	kilometres (km)
Weight or mass	grams (g)
	kilograms (kg)
	tonnes (t)
Liquid	millilitres (ml)
	litres (l)
Area	square metres (m²)
Volume	cubic metres (m³)

Metric units Decimal units of measurement based on the metre and the kilogram

Exam tip

Make sure you are clear about how to move between different units, for example:
+ 1 mm × 10 = 1 cm
+ 1 mm × 1000 = 1 m
+ 1 m × 1000 = 1 km.

Moving the decimal point changes the value, so:
+ 6250 mm can be shown as 6.250 m
+ 6250 m can be shown as 6.250 km.

> **Worked example**
>
> How many standard bricks are required for a half-brick thick wall 3.5 m long and 1125 mm high?
>
> Formula: area (m²) = height (m) × length (m)
>
> Convert 1125 mm to m = 1.125 m
>
> 1.125 m (height) × 3.5 m (length) = 3.9375 m²
>
> = 3.94 m² to two decimal places
>
> Formula: 60 bricks per m² for a half-brick thick wall:
>
> 3.94 × 60 = 236.4 bricks = 237 rounded to whole bricks

> **Exam tip**
>
> Exams often feature calculation questions using area and volume. Memorise the formulae for these calculations. Get into the habit of showing your calculations in a logical sequence.

Measuring pressure

Every building must be designed to withstand the loads and pressures created by:

+ the weight of the materials it is constructed from (known as dead loads)
+ the weight of equipment/items and occupants within it (known as live or imposed loads)
+ external forces, such as wind or the weight of snow (also known as live or imposed loads).

These forces are measured in newtons (N). For example, a crushing force exerted on concrete can be measured as newtons per square millimetre (N/mm²).

> **Exam tip**
>
> When answering questions about weight and mass, it is important to understand the difference between live (or imposed) loads and dead loads.

Measurement techniques

REVISED ●

In construction, standard units of measurement are used in specific ways using appropriate equipment.

Simple measuring equipment

Simple equipment such as a tape measure can give acceptably accurate results when used carefully to measure:

+ height (for example of floor positions and ceilings, door and window opening heights and roof heights)
+ length (for example the outline of the building, position of internal walls and the position of doors and windows along an elevation of a building)
+ distance.

> **Exam tip**
>
> If a question on measurement includes drawings or diagrams, take the time to study them carefully until you fully understand what is required.

Complex measuring equipment
Measuring height, length or distance

Technically complex equipment can be used to measure height, length or distance to high levels of accuracy:

+ Laser instruments can be used to measure differences in height.
+ 3D laser scanning can be used to give accurate data about the height and length of existing structures during a survey.
+ Traditionally, an instrument called a theodolite has been used to measure distance and height using mathematical methods of trigonometry and triangulation to establish accurate angles and measurement details for setting out new buildings or surveying existing structures.
+ The optical theodolite has evolved into the 'total station', which is an electronic version of the theodolite. This can record location data using GPS, with a distance meter to speed up the measurement process.

> **Elevation** View of the front, back or sides of a building
>
> **Trigonometry** Branch of mathematics concerned with relationships between angles and ratios of lengths
>
> **Triangulation** Surveying method that measures the angles in a triangle formed by three survey control points

Measuring weight or mass

There is usually no means of weighing things on site, but the weight of items should be labelled on packaging, and equipment for lifting and transporting materials should have a safe working load (SWL) clearly and permanently indicated on the machinery.

Data sources

Calculations using units of measurement for a range of applications can be undertaken from different data sources. Traditional data sources such as drawings and specifications are commonly used to 'take off' information for calculating quantities of materials and establishing timescales for work programmes.

Use of ratios

REVISED

When calculating quantities, ratios are often used to establish material requirements. For example, a specification may require concrete to be mixed using 5 parts of aggregate or chippings, 3 parts of sand and 1 part of cement, expressed as a ratio of 5:3:1.

There is more below on the use of ratios in describing scale measurements.

Now test yourself TESTED

5 Explain why a metric system of measurement is easy to use.

6 How is 1000 millimetres converted to metres?

7 Outline **two** possible consequences of inaccurate measurement in relation to height, length and distance.

8 What is a total station used for?

9 Explain the difference between U-values and R-values.

Exam tip

Make sure you are familiar with the range of traditional data source documents, such as schedules and bills of quantities, besides the specification. Understand the purpose of each one.

6.3 Measurement standards, guidance and practice

Conventions

REVISED

Construction drawings are produced to a set of conventions. These include:

+ specific units of measurement
+ specific views and viewing angles
+ specific sheet sizes.

Conventions Agreed, consistent standards and rules

Scale

REVISED

An architectural technician or a draughtsperson will produce drawings of a building to scale.

Scale is shown by using a ratio such as 1 to 10, which is usually written in the form 1:10.

Using a scale of 1:10, a drawing of a feature that is 1m (or 1000mm) long in real life would be drawn 100mm (or 10cm) long. This is because 100mm is one tenth of 1000mm.

In some cases, more than one scale will be used on the same drawing, with different scales being used for different views, perhaps to allow for greater detail to be shown or a wider view to be given in context.

Scale When accurate sizes of an object are reduced or enlarged by a stated amount

Common scales

REVISED

Table 6.3 shows the scales commonly used for different types of drawing.

Table 6.3 Types of drawing and common scales

Type of drawing	Description	Scales commonly used
Detail drawing	This is a very accurate, large-scale drawing of the construction of a particular item or part of a building.	1:1, 1:2, 1:5 and 1:10
Floor plan	This shows the layout of internal walls, doors, stairs and, in a dwelling, the arrangement of special-use rooms such as bathrooms and kitchens.	1:50 and 1:100
Elevation drawing	This shows the external appearance of each face of the building, with features such as slope of the land, doors, windows and the roof arrangement.	1:50 and 1:100
Sectional drawing	This is a slice or cut through of a structure to give a clear view of details that would otherwise be hidden.	1:50 and 1:100
General arrangement drawing	Sometimes referred to as a location drawing, this can be used to show a single building element and what it should contain. It can also be used to show the main elements of a structure in context, such as the external walls, internal or partition walls, floor details and stairs.	1:50, 1:100 and 1:200, depending on the level of detail required
Site plan	This shows: ✚ the proposed development in relation to the property boundary ✚ the positions of drainage and other services ✚ access roads and drives ✚ the position of trees and shrubs if they are required as part of the planning details.	1:200 or 1:500
Block plan	This shows the proposed development in relation to surrounding properties. It usually shows individual plots and road layouts on the site as a simple outline with few dimensions.	1:1250 or 1:2500

British Standard requirements

As information sources that can be reliably used during both the planning and construction stages of a project, drawings should comply with British Standard requirements.

Until 2019, BS 1192 used to set out quality standards and methods of managing the production and distribution of construction information, including drawings. This has now been replaced by BS EN ISO 19650.

To produce drawings that are consistent and can be formatted to suit particular applications, they are laid out on standard-sized sheets of paper.

Table 6.4 shows the dimensions of standard paper sizes. Note that the higher the number, the smaller the paper.

Table 6.4 Standard paper sizes

Title	Size (mm)
A0	1189 × 841
A1	841 × 594
A2	594 × 420
A3	420 × 297
A4	297 × 210

Rules of measurement

Costing a project from drawings and other data sources must be done in accordance with 'rules of measurement'. This is a system of standardised methods that ensures consistency, helping to avoid disputes and maintain efficiency and productivity.

The rules are contained in a publication called New Rules of Measurement (NRM).

Tolerances

REVISED

Tolerances are necessary because:
+ a building will often be constructed in conditions that are not perfect
+ many materials have characteristics that make it difficult to produce a flawless result.

A tolerance is expressed as plus or minus (±) a stated amount. Tolerances will vary, depending on the nature of the material.

Tolerances may also be established to allow for:
+ acceptable variations in the strength of materials
+ temperature ranges in which materials can be used.

Exceeding allowable tolerances for a given construction task will mean that the work does not meet the specification.

Tolerances Allowable variations between specified measurements and actual measurements

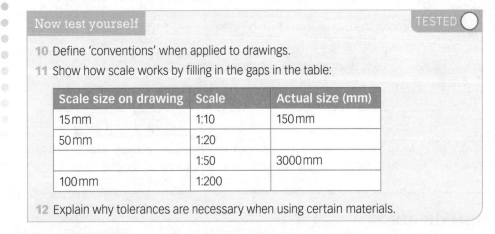

Now test yourself   TESTED

10 Define 'conventions' when applied to drawings.

11 Show how scale works by filling in the gaps in the table:

Scale size on drawing	Scale	Actual size (mm)
15 mm	1:10	150 mm
50 mm	1:20	
	1:50	3000 mm
100 mm	1:200	

12 Explain why tolerances are necessary when using certain materials.

Exam checklist

In this content area, you learned about the following:
+ accurate and appropriate measurement
+ standard units of measurement and measurement techniques
+ measurement standards, guidance and practice.

Check your understanding and progress at **www.hoddereducation.co.uk/myrevisionnotes**

Exam-style questions

Short-answer questions

1 Explain how accurate measurement can contribute to creating a realistic work programme. [2]

2 State **two** potential consequences if a building is not accurately set out to specified measurements. [2]

3 Explain why accurate linear measurement is important when locating existing services on a new project site. [1]

4 Describe the characteristics of job costing. [4]

5 State **two** ways in which life-cycle cost analysis can benefit the environment over time. [2]

6 List **two** construction details that can be shown on a floor plan. [2]

7 Which type of construction drawing is produced to a scale of 1:1250 or 1:2500? [1]

8 State **two** specific factors that are included in the conventions applied to producing drawings. [2]

9 List **two** activities for which a laser instrument can be used to take accurate measurements. [2]

10 Discuss reasons why a laser distance measurement tool is more accurate than a tape measure. [6]

11 How are the forces created by weight bearing on a structure measured? [2]

12 Give a definition of 'tolerances' and give examples of how they are used when measuring and calculating quantities. [4]

Extended-response questions

13 A detached house measuring 10m by 8.5m is to be built centrally on a plot measuring 50m by 60m. Discuss the range of measurements that will be used to:
   + establish the footprint of the house
   + strip 200mm depth of topsoil
   + position and excavate trenches for foundations and drains
   + locate existing services. [9]

14 Discuss how activity costing could be used to establish the cost of constructing a 1.2m high boundary wall around the perimeter of the plot described in the previous question. [12]

7 Building technology principles

7.1 Construction methods

Traditional and modern construction methods are summarised in Table 7.1.

Table 7.1 Traditional and modern construction methods

Period	Construction methods
Georgian (1714–1837)	+ Single-skin stone or handmade brick walls + Vertical sliding wooden sash windows + Symmetrical facades
Victorian (1837–1901)	+ Single-skin brick solid walls, built on concrete, hydraulic lime or stepped brick foundations + First use of cavity walls + First use of a damp-proof course (DPC) + Pointed arches constructed over doors and wood- or metal-framed windows
Edwardian (1901–10)	+ External cavity walls + Electric lighting + Some timber framing + Hanging wall tiles + Pebbledash walls + Timber porches and balconies
Addison Act (1919)	+ Introduction of affordable council housing after the First World War + Commonly three to four bedrooms, indoor toilets, baths and hot running water + Often constructed with brick, block or concrete walls + Window openings designed for natural light
Art Deco (1920s–40s)	+ Flat roofs + Metal-framed ('Crittall') windows + Open interiors with Egyptian influences + Other forms of foundations introduced, including piles and raft
Semi-detached (1930s)	+ Hipped roofs and pebbledash walls + Single-skin brick walls bonded with lime mortar + Recessed porches + Mock timber framing + Timber bay windows
Prefabs (1940s)	+ Mass produced in factories and assembled on site + Precast concrete columns and metal tubing + Small windows
Terraced (1960s–70s)	+ Clad with hanging wall tiles or weatherboarding + Single-glazed aluminium windows and doors + Polythene damp-proof membrane (DPM) used as a barrier to prevent moisture entering the building
New build (1990s)	+ uPVC double glazing + Insulated roofs, floors and cavity walls with a rendered finish
Current day	+ Energy efficient, eco-friendly building materials + Open plan + Use of glass for solar gain

Onsite construction methods

REVISED

In onsite construction, most of the building components are erected or assembled by skilled workers, from the foundations to the roof. When using traditional methods such as brick and block cavity walls or timber frame, progress can be slow.

Check your understanding and progress at **www.hoddereducation.co.uk/myrevisionnotes**

Alternative methods considered less harmful to the environment include straw-bale walls and, less conventionally, shipping containers. 3D printing and robotics are also used as methods of onsite construction.

Progress with this method of construction can be delayed by adverse weather conditions.

First fix

REVISED

This term has a different meaning for each trade, although it is generally considered to be the phase of work completed after the structure of a building has been erected and before plastering commences. Work undertaken might include installing:
+ water pipes
+ electric cables and back boxes
+ pipes for gas distribution and ducting for heating, ventilation and air-conditioning systems.

Second fix

REVISED

This stage of construction happens after first fix and plastering/drylining. It involves the installation of items and equipment that could have been damaged or affected by earlier stages of construction work. Work undertaken might include installing:
+ plumbing fixtures and fittings
+ electrical fixtures and fittings, and testing and commissioning new electrical systems.

Off-site construction methods

REVISED

This system of construction combines pre-engineered units (or modules) to form major elements of a building. These modules are manufactured in factories and then transported to site, where they are assembled.

Off-site construction methods include pre-assembled, precast, modular, panel systems and 3D printing.

Self-driving vehicles

REVISED

Using autonomous (self-driving) vehicles for construction projects can reduce costs by improving productivity and site progress. This is because they can operate 24 hours a day without a driver.

Computer-controlled manufacturing robots

REVISED

These are designed to improve productivity by undertaking repetitive and labour-intensive tasks, such as demolition, laying bricks and blocks, and plastering.

Large-scale 3D printers

REVISED

These can produce perfectly formed walls for large buildings on demand, using a robotic arm onsite to 'print' cement-based mortar walls, layer by layer.

Drones

REVISED

This technology is used in the construction industry to plan, manage, report and communicate efficiently through digital platforms, for example BIM. Drones can reduce costs and labour needs and improve the efficiency of construction projects.

With the right software package, drones can be used for the following purposes:

+ land surveying
+ producing 2D and 3D maps
+ identifying improvements that can be made to infrastructures
+ pre-construction site planning
+ maintenance inspections
+ thermal imaging
+ streaming live footage to the project team with the use of Virtual Reality (VR) glasses
+ inspecting work for quality control
+ monitoring progress on site
+ identifying hazards and improving site security.

Refurbishment and renovation

Refurbishment work in the construction industry is focused on cosmetic repairs to a building or structure to improve its condition.

Renovation work is more intrusive, and often involves changes to the structure of an existing building.

> **Renovation** Alterations or improvements made to the fabric or structure of the interior or exterior of a building

Maintenance

Throughout the lifetime of a building, its fabric and building services will have to be maintained in order to:

+ keep them working efficiently
+ prevent the building from deteriorating.

As new technologies and materials are developed, the building may be improved with retrofitted upgrades to its sources of energy and insulation.

Exam tip

Make sure you know about the different types of construction method. You may be asked questions about the advantages of off-site construction and energy efficiency, for example.

Typical mistake

Many students make the mistake of thinking off-site construction methods are typically slower than onsite manufacturing. Manufacturing off site is quicker than onsite because it is more efficient and there are no delays caused by adverse weather conditions.

Now test yourself

 TESTED

1 At what stage does second fix take place in onsite construction?
2 What are the advantages of using large-scale 3D printers to construct walls on construction sites?
3 What are the advantages of off-site construction?

Revision activity

Create a list of traditional methods of wall construction. Explain the disadvantages of each method and give a modern alternative.

7.2 Forms of construction

In the construction industry, the structure of a building is divided into two parts, as illustrated in Figure 7.1.

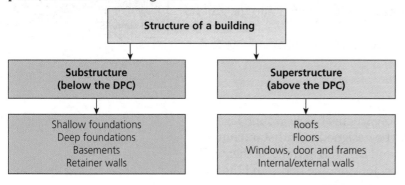

Figure 7.1 The structure of a building

Substructure

Shallow foundations

There are several types of shallow foundation commonly used to construct buildings up to four storeys high, and these are described in Table 7.2.

Table 7.2 Shallow foundations

Shallow foundation type	Key features	Benefits

Narrow strip and trench fill foundations (also known as strip foundations)	+ Narrow trench approximately 600 mm wide and the length of load-bearing walls
	+ Minimum widths for each ground type in Approved Document A
	+ Usually at least 1 m deep, or until load-bearing soil is reached
		+ Depth avoids frost damage
		+ Trench fill foundations:
		  + generally cheaper and safer to build than traditional strip foundations
		  + quicker to construct
		  + avoid the risk of the sides of the trench collapsing

Wide strip foundations	+ Similar in construction to narrow strip foundations + Typically at least 1.5 m deep and much wider + Distribute the weight through the reinforced concrete foundation over a wider area	+ Provide enough space for bricklayers to work safely in trenches to build walls to the DPC level + Support superstructure of buildings with heavier loads or where soil has lower bearing capacity
Raft foundations	+ Used for small, low-rise domestic buildings + Concrete slab reinforced with steel frame is set under entire building + Slab acts like a raft on water, spreading loads over a large area + Where additional loads occur, e.g. load-bearing walls, foundation depth is increased	+ Relatively quick to construct + Use less concrete compared with other methods
Pad foundations	+ Used to construct industrial units and commercial buildings + Constructed around a steel portal frame, with suspended external cladding and low-height walls + Most weight is transferred from the superstructure through the steel frame to single point loads at ground level, where square or circular concrete pad foundations distribute the loads	+ Support quick and efficient construction of buildings + Can also be used to support ground or ring beams

95

Deep foundations

Deep foundations support the weight of high-rise buildings from the bedrock below the ground. Alternatively, they rely on friction in firm soil.

There are several types, including:
+ driven piles – piles that are pressed, vibrated or hammered into the ground with construction plant/machinery
+ replacement piles – holes bored into the ground, filled with concrete and reinforced with steel.

Typical mistake

Students often lose marks because they have misunderstood the purpose of different types of foundations.

Basements

Basements are habitable rooms or spaces constructed below ground level.
+ The walls and floor of a basement must be designed to withstand imposed loads from the ground enclosing them, and the weight from further walls and other elements of the building above, for example floors and the roof.
+ Parts of a structure that are below ground level, such as a basement, are at risk of moisture ingress and flooding, therefore particular care must be taken in their design and construction to keep the building dry.

Retainer wall

A retainer wall is a structural wall designed to bear the weight of lateral loads imposed from one side, either above or below ground level.

Figure 7.2 Retainer wall

Superstructure

REVISED

Roofs

On domestic buildings, roofs are either flat or pitched:
+ Flat roofs have a slight fall up to 10 degrees and can be covered with metal, fibreglass, glass, rubber or a green roof.
+ Pitched roofs have a fall above 10 degrees and can be covered with tiles, slates, metal, fibreglass or solar tiles.

To protect a building from heat loss, roofs are either insulated between the rafters (known as a cold roof) or over the rafters (known as a warm roof).

Exam tip

Relevant maths questions could be included in your exam. For example, you could be asked to calculate the length of a rafter on a roof using Pythagoras' theorem $a^2 + b^2 = c^2$, as shown in Figure 7.3.

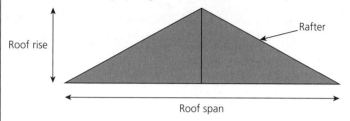
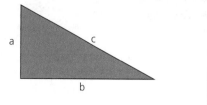

Figure 7.3 Using Pythagoras' theorem to calculate rafter length on a roof

Floors

There are two types of floor construction:

+ Solid floors (see Figure 7.4) are constructed using different layers to make a solid concrete base.
+ Suspended floors are hollow floors constructed with timber and timber-based joists, or concrete blocks and beams.

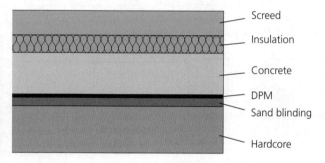
Screed
Insulation
Concrete
DPM
Sand blinding
Hardcore

Figure 7.4 Solid floor

Windows, doors and frames

High-performance, energy-efficient doors, windows and frames can be made from hardwood, softwood, uPVC or aluminium.

The type of glass used is just as important as the frames. Double and triple glazing is mainly specified for new construction projects because the high-performance, low-emissivity (low-E) glass and argon-gas-filled units act as insulators to reflect heat back into a building to reduce energy bills.

Internal/external walls

+ Cavity walls are used for external walls because insulation can be sandwiched between the internal and external skins to reduce heat loss.
+ Structural timber frame buildings are usually infilled with various types of insulation.
+ In modern construction, solid walls are only used for external work, or load-bearing internal walls.
+ Non-load-bearing internal partition walls are often made from timber or metal studs.
+ Vertical and horizontal damp proofing is used at ground level and where there are door and window openings to prevent water ingress.
+ Metal, stone, timber or concrete lintels are used to bridge the gap over the openings in a wall created for the fenestration of a building.

Infrastructure

Roads, sewage systems, railways and bridges are examples of infrastructure that may be altered or constructed as part of a building project.

External work

External construction work could involve the following areas:

+ paving
+ boundaries
+ drainage
+ parking.

Most roads and pavements in the UK are constructed with a sub-base of compacted coarse aggregates and covered with layers of durable asphalt, also known as tarmac. Asphalt is durable and quick to lay, making it a cost-effective alternative to other materials such as:

+ concrete
+ block paving (permeable surface)
+ gravel (permeable surface)
+ resin-bound gravel (permeable surface).

Special consideration must be taken with regards to drainage on driveways and roads to comply with planning permission and to prevent flooding from rainwater runoff.

> **Permeable** Porous, allowing water to drain through

> **Exam tip**
>
> There is a lot to learn in this section, so try to think of a building in different parts as if you are constructing it from the foundations to the roof, with all the elements in between.

> **Now test yourself**   TESTED ○
>
> 4   Name **two** examples of deep foundations.
> 5   Which building component is used to bridge the gap over the openings in a wall created for the fenestration of a building?

> **Revision activity**
>
> Research different types of soil and suggest suitable foundations for a building constructed on these ground conditions.

7.3 UK building regulations and approved documents

Building regulations   REVISED ○

The purpose of building regulations is to define the minimum standards of design, building materials and work in the UK.

Virtually all renovations and construction of buildings and building services must comply with building regulations. If you are in doubt, you must check with building control before constructing or changing buildings in certain ways.

Approved documents   REVISED ○

The Ministry of Housing, Communities and Local Government publishes a range of approved documents to help people understand and comply with building regulations:

+ A: Structure
+ B: Fire safety
+ C: Site preparation and resistance to contaminants and moisture
+ D: Toxic substances
+ E: Resistance to the passage of sound
+ F: Ventilation

- G: Sanitation, hot water safety and water efficiency
- H: Drainage and waste disposal
- J: Combustion appliances and fuel storage systems
- K: Protection from falling, collision and impact
- L: Conservation of fuel and power
- M: Access to and use of buildings
- P: Electrical safety
- Q: Security – dwellings
- R: Physical infrastructure for high-speed electronic communication networks

Exam tip

Dedicate some time to researching the purpose of building regulations and understanding each of the approved documents.

Typical mistake

Students sometimes get confused between planning permission and building regulations. Planning permission is the approval needed from the local planning department before most building work can start; building regulations are the standards that all building work in the UK must comply with.

Now test yourself   TESTED ◯

6 What supportive information is published by the Ministry of Housing, Communities and Local Government to help people understand and comply with building regulations?

7.4 Building standards

The British Standards Institute (BSI) produces agreed standards across a wide variety of industry sectors. In construction, these standards relate to structures, materials and sustainability. The purpose of British Standards is to provide general and specific guidance for the construction industry to improve working practices.

The BSI also develops Publicly Available Specifications (PAS). These are fast-tracked standardisation documents produced to meet urgent market need.

The International Organization for Standardization (ISO) is an independent, non-governmental organisation that develops and publishes international standards.

The Common Minimum Standards for Construction (CMS) set out mandatory standards that construction project team members in government should use for projects in the built environment. The purpose of the CMS is to ensure procurement decisions give the best value for money for taxpayers and help meet the government's objectives, for example, the development of skills and sustainability.

Exam tip

Acronyms are often used for building standards, so make sure you are familiar with the following: BSI (British Standards Institute), BIM (Building Information Model), PAS (Publicly Available Specification), ISO (International Organization for Standardization).

International Organization for Standardization (ISO)
Independent, non-governmental organisation that develops and publishes international standards

Revision activity

Create flashcards with the acronyms used in this section on one side and their full meaning on the other. Place the cards face down with the acronyms on top. Practise selecting a card and trying to remember the acronym's meaning before turning it over.

7.5 Manufacturers' instructions

The purpose of manufacturers' instructions

Any product that has been designed and manufactured in the UK, or imported into the UK, must conform to section 6 of the Health and Safety at Work etc. Act 1974 (HASAWA) – *General duties of manufacturers etc. as regards articles and substances for use at work.*

This legislation states that:
+ products must be designed and constructed to eliminate risks, so that they are safe while being used, cleaned and maintained
+ where necessary, products must be calibrated, tested, inspected and certified to meet with relevant product regulations
+ adequate information must be provided for the safe maintenance, operation and disposal of products
+ as far as is reasonably practicable, revisions of information must be supplied if a serious risk to health or safety concerning a product has been discovered.

Types of manufacturers' instructions

Product information is usually supplied with goods in the form of printed manufacturer's instructions, with further copies available on the manufacturer's website.

Installation instructions

Where applicable, installation instructions provide competent installers with the minimum industry standards expected to comply with building regulations for the safe fitting of a specific product.

Products that are not installed as the manufacturer intended may be hazardous, therefore invalidating any manufacturer guarantee.

Operation instructions

Operation instructions inform installers, end users and those who maintain or service products how to use a product, while not compromising the health and safety of themselves or others. Products that are used as the manufacturer intended will often be more efficient and last longer.

Maintenance instructions

Manufacturers often provide recommended times for routine maintenance and servicing of their products, and details of how this should be completed. They may also provide details on common faults and how these may be rectified by a competent person.

> **Now test yourself** TESTED
>
> 7 Under which legislation do manufacturers have to provide specific product information?

> **Revision activity**
>
> Source four different types of manufacturers' instructions. Analyse the instructions and make a list of the similarities between them.

> **Exam tip**
>
> There are requirements under health and safety legislation for manufacturers to provide information on their products. Revisit Chapter 1 to recap the Health and Safety at Work etc. Act 1974 (HASAWA).

Exam-style questions

Short-answer questions

1 Describe the superstructure of a building. [1]

2 Explain why a DPC is used in the construction of cavity walls. [1]

3 List the tasks that a drone can be used for in the construction industry. [2]

4 Explain why windows, doors and frames are subject to planning permission and inspection by building control after they have been installed. [2]

5 Identify **two** types of floor construction. [2]

6 Identify the type of work undertaken by a plumber at the second-fix stage of building work. [2]

7 A contractor excavates a narrow trench for a strip foundation. The building inspector inspects the trench and is unhappy with the ground conditions for this type of foundation. Identify an alternative foundation for this project. [3]

8 Describe modular construction. [3]

9 List the components used to construct a solid concrete floor at ground level, in the correct order from the ground up. Note – the floor does not contain underfloor heating. [3]

10 Explain the difference between a warm roof and a cold roof. [3]

11 Calculate the (true) length of a rafter if a double-pitched roof has a rise of 1.8 m and a span of 4.6 m. Show your calculations. [3]

12 List the types of shallow foundation used in the construction industry. [4]

13 Explain the difference between the first-fix and second-fix stages of building work. [4]

14 Describe the benefits of modern construction methods. [4]

15 State **five** advantages of using autonomous vehicles for construction work. [5]

Extended-response questions

16 Analyse the use of a range of different products and materials that could be used to provide a waterproof covering for various shaped pitched roofs on domestic properties. [9]

17 Evaluate the benefits of portal-frame construction for industrial buildings. [12]

8 Construction information and data principles

8.1 Data

Data is used in planning, costing, constructing and operating a building.

+ Analysis and evaluation of data can allow informed decisions to be made when bringing a project to a successful completion.
+ Projections can be made about the efficient operation of a building throughout its life cycle.

Working with data has developed into a science that follows defined methods to perform specific tasks.

Information and data are closely linked. Data is a collection of facts from which meaningful information can be extracted. Examples of data include:

+ numbers
+ measurements
+ words or descriptions.

> **Exam tip**
>
> To develop the right mindset about data, whenever you use a search engine to find specific information, think about what type of data you are presented with in the search results. Is it numbers, measurements, words or descriptions? Or is it some other factual format that information can be extracted from, such as a photograph?

Key elements of data

REVISED ◯

Vast quantities of data are generated each day. This is often referred to as 'big data'.

Accuracy, interoperability and metadata

Simply collecting masses of data is not the key to creating useful information. To be useful, data must be:

+ 'cleaned' to remove inaccurate or imprecise data
+ identified to assign it for use
+ stored correctly.

Managing data in this way can involve many complex data-processing systems working together, requiring them to be operationally compatible. This is known as data interoperability.

Complex processing of data can involve one system describing or analysing data in another system. The set of data produced by this process is called metadata.

Level of detail

The level of detail in a set of data (or dataset) may need to be regulated or even limited, so that the type of information it provides is clear.

Generalisation

To allow the extraction of useful information from large datasets, a process known as generalisation can be used. This is a way of taking a broader view of the mass of raw data to provide a more general view.

> **Data interoperability** Ability of data systems to exchange and use information
>
> **Metadata** Set of data that describes or analyses other data
>
> **Generalisation** (In data processing) creating layers of summarised information from a mass of data

Check your understanding and progress at **www.hoddereducation.co.uk/myrevisionnotes**

This reduces the volume of data while still providing useful information, to allow analysis of factors such as heating and ventilation requirements or optimal occupancy levels of the completed building.

Revision activity

An example of how metadata works can be seen in the use of spreadsheets. Find an online spreadsheet and experiment with numerical data entries to display and manipulate numbers.

Different sources from which data can be generated

REVISED

Data generated during design

When a building is designed, data will be generated regarding:
+ overall dimensions and specific measurements of the structure
+ floor area and volume of rooms
+ energy requirements
+ thermal performance of the structure
+ structural strength and the forces generated by loadings.

This data will be used in different ways by each member of the design and construction team, such as:
+ quantity surveyors preparing costings
+ contractors and subcontractors preparing tenders
+ systems designers planning building services requirements.

Building services systems require specific data to establish dynamic elements, such as:
+ air flow and gas supply rates for heating and ventilation systems
+ water flow rate and pressure calculations
+ electricity voltage and current ratings.

Data generated during construction

During the construction phase, data analysis will be used to:
+ monitor progress towards deadlines
+ manage efficient interactive working between site personnel
+ monitor the speed of operations and the effects of delays caused through bad weather or supply issues.

Building Information Modelling (BIM)

Building Information Modelling (BIM) is increasingly used to integrate many data sources in a digital format. BIM allows access to a range of important information by authorised users during the design, construction and occupation of a building, right through to demolition of a structure. See page 106 for a more detailed discussion of BIM.

Infrastructure and transport systems

Within the built environment, data is used to design, construct and maintain vital infrastructure and transport systems.

Data generated on completion and handover of a building

When a building is completed and handed over from the contractor to the client for occupation and brought into use, analysis of data can be used to answer questions such as:
+ How successful was the delivery of the project by the contractor?
+ Was the project delivered and handed over on time? If not, why not?
+ Does the building meet its design brief fully now it is in use?
+ What changes could be made to improve performance?

Exam tip

Identify the type of data that applies to each phase in a project. For example, the design phase will generate numerical data in the form of measurements, the construction phase will generate ordinal data in the form of planned sequences of operations, etc.

Loadings Application of a mechanical load or force on a structure

Dynamic Characterised by frequent change or motion

Exam tip

You could be asked a data-related question about any of the phases of a building project. Before answering, try to imagine yourself as one of the personnel who would work on the identified phase. For example, during the design phase, consider how quantity surveyors would use and create numerical data.

Revision activity

Buses and trains are an important part of the UK's infrastructure. Make a list of information that you think would be needed to create bus and train timetables. What ongoing data would need to be gathered to ensure the timetables are suitable for travellers' needs?

Post-occupancy evaluation (POE)

Gathering data to answer questions such as those listed on the previous page is referred to as post-occupancy evaluation (POE). It can provide valuable data to contractors who build repeat structures or specialise in specific project types.

In most buildings, there are components that automatically control various systems such as heating thermostats or alarm motion sensors. In large structures, sophisticated computer-based building management systems can be installed to automatically control and monitor aspects such as:

+ general energy consumption and distribution
+ lighting levels
+ water management and consumption rates.

The data generated can be utilised to:

+ understand how buildings are operating
+ adjust and control systems to optimise their performance
+ create reports
+ set alarms to give an alert when operational parameters are exceeded.

Enterprise systems

In the case of large estate facilities, such as a hospital complex or university campus, specialised computer software referred to as enterprise asset management (EAM) can be employed to manage the flow of data.

Data generated during maintenance

EAM can be used to collect data about efficiency, reliability and repair costs, allowing planning and work scheduling for maintenance.

As inspections take place, data can be logged in real time, and work orders can be sent to personnel already on site to prioritise maintenance or repair tasks for completion.

Data is generated, processed and stored using information and communications technology (ICT).

Ways data can be used

REVISED

Data is a key part of the construction process, from design to demolition:

+ Data networks that transfer information efficiently between users have become integral to the successful completion of construction projects.
+ Accurate information derived from reliable and accessible data sources is essential in determining costs and maintaining profitability across all areas of construction activity.

Table 8.1 identifies some ways in which data can be used.

Table 8.1 Ways in which data can be used

Data uses	Applications
Understanding behaviour	Data can be used to analyse and understand the behaviour of personnel working on a project.
	Efficient deployment of skilled workers and the creation of workforce motivation to achieve operational efficiency are valuable management skills. Data regarding personnel numbers on site can be referenced to output over time to establish optimum levels of staffing.
	With the emergence of new materials and the progressive adoption of new construction methods, reference to 'behaviour' could be extended to factors that might impact on project success, such as behaviour of certain materials and equipment in variable conditions.
Performance assessment	**Key performance indicators (KPIs)** can be linked to appropriate data to measure performance in areas such as: + monitoring project costs + tracking project progress over time + identifying company strengths and weaknesses + confirming client satisfaction.

> **Exam tip**
>
> When writing your answers, use the correct terminology to receive high marks. Terms such as 'interoperability' and 'metadata' are technical terms that have important applications when discussing data and information.

> **Key performance indicator (KPI)** An area within a business that can be measured in terms of a value, for example project completion times

Check your understanding and progress at **www.hoddereducation.co.uk/myrevisionnotes**

Data uses	Applications
	Uses for performance data include: + tracking profitability by comparing cost with budget + highlighting trends in complaints from clients + evaluating quantities of waste with a view to improving recycling.
Improving market competitiveness	When companies bid for contracts, accurate data can be used to ensure they are well informed about: + workforce availability + materials and components procurement + current regulatory requirements + fluctuating economic conditions. Accurate and current data is critical in: + providing reliable information, so that a company can improve and maintain its success + fully understanding the area of construction activity that a company works in, to inform decisions regarding investment in training and equipment needs + assessment of which types of construction project match the capabilities and experience of a company, to maintain profitability.
Allocation of resources	A company may work on several projects simultaneously. This means that assets and resources will need to be actively managed to make best use of them across the range of work being progressed. Efficient analysis of resource allocation relies on accurate data concerning: + personnel placement + work activities + plant and equipment usage + temporary accommodation on site.

Exam tip

Study the tables in revision guides carefully. They separate important information into more easily remembered sections, so make good use of them. Writing your own summary tables is a good memory aid.

Revision activity

Look at the 'Data uses' column in Table 8.1. Which of the four uses identified would be applied every week on a construction project? Write a paragraph explaining your choice.

Now test yourself

TESTED

1 Why does data need to be processed to be useful?
2 What kinds of data can be generated during the design stage of a building project?
3 When is BIM used to store and analyse data?
4 List the benefits of data analysis in each phase of a building's life cycle.
5 How can data be used to improve a company's market competitiveness?

8.2 Sources of information

Interpreting data sources

REVISED

Specific data sources may provide information for workers with designated responsibilities.

Specific sources of information can be categorised and the personnel groups or individuals that use them can be identified. Information and data sources used within construction and building services projects include:
+ product data
+ manufacturers' specifications
+ client's specifications
+ Building Information Modelling (BIM)
    + Common Data Environment (CDE)
+ work programme planning
+ commissioning
+ certification
+ test data schedules
    + condition reports
    + emissions testing.

Product data

This is information or instructions about how to correctly use or install a product. A manufacturer must supply data or information about how to handle and store a product safely in accordance with health and safety legislation, such as the Control of Substances Hazardous to Health (COSHH) Regulations 2002.

As an information source, product data can be used by:
+ onsite personnel installing a range of systems
+ trade workers confirming that materials are in accordance with specifications for the work task.

See section 1.12 for examples of symbols which could be used as product data.

Manufacturers' specifications

Product data is closely tied to manufacturers' specifications, which give very specific information on performance data.

Manufacturers' specifications may include key data and information on how a product or piece of equipment should be:
+ assembled or dismantled
+ maintained or repaired
+ examined or inspected after installation.

Client's specification

In the Royal Institute of British Architects' (RIBA) Plan of Work 2020, the client's specification has been defined in part as 'a statement or document that defines the project outcomes and sets out what the client is trying to achieve'.

Information within the client's specification must be interpreted carefully to produce a design brief that will meet the objectives as fully as possible.

Building Information Modelling (BIM)

Most onsite construction operatives will not use BIM directly, but an awareness of what it is and how it works is an advantage.

BIM is a structured system that allows access to a range of important information by authorised users at all stages of construction, handover and occupation of a building. It makes use of digitally processed information to analyse design elements of a building, including 3D modelling.

Using BIM, complex design ideas can be transformed into a data medium that is potentially easier for all personnel to work with.

BIM allows digital data describing internal building engineering services of a structure to be presented and analysed visually, so that the way they interact with each other can be seen. Identification of clashes in systems (referred to as 'clash detection') can then be revealed at an early stage in the design process.

Common Data Environment (CDE)

The Common Data Environment (CDE) is a single central source of information used within the BIM system. All relevant documents, information and other data sources are brought together in a digital environment that can be accessed by all authorised personnel collaborating on the project.

The CDE can contain different types of digital information, including:
+ schedules
+ contracts
+ registers
+ reports
+ 2D drawings
+ 3D models.

> **Exam tip**
>
> You may have to answer a question on product data for any construction materials or equipment. You can help prepare for this by habitually examining product information labels on items you come across and becoming familiar with the sort of data they provide.

> **Typical mistake**
>
> Students sometimes lose marks for describing BIM as simply a 3D model of a project. BIM is much more than that – it brings together an array of digital information sources that usually include a 3D model.

> **Exam tip**
>
> Questions about BIM will focus on what it does and how it does it, rather than the individual names of all its elements (such as the CDE). Unless you are a BIM specialist, you are unlikely to be familiar with the vast range of BIM labels and descriptors.

Information in the CDE forms the foundation of shared information on which collaboration can take place, making it possible to reduce mistakes and avoid duplication.

Work programme planning

To ensure work can be completed on schedule and within budget, careful prior planning of the construction process is required. Planning a programme of work is often undertaken using charts that give data and information about the sequence of planned activities.

There are two main documentary methods of planning the sequence of work in construction:
+ Gantt charts
+ critical path analysis (CPA).

Commissioning

Commissioning is the process of using and interpreting data to confirm that all building systems are installed, tested, operated and maintained according to predetermined operational requirements.

There are circumstances where performance testing of systems may continue after the client has accepted handover. For example, some systems are dependent on different weather conditions, and commissioning inspections and adjustments may take place over the first year of occupation.

Certification

Following commissioning, a certificate is issued to confirm that installation and operational standards have been met.

Certification must comply with technical benchmarks and criteria set by standards organisations.

Test data schedules

Even after a building's systems have been commissioned and certified, ongoing inspection and testing are required throughout the life cycle of the building; for example, heating systems that use gas or oil burners will need testing to ensure carbon emissions and harmful gases are within specified limits. A schedule for this testing will include data that needs to be recorded to ensure optimal operational efficiency is maintained.

Records of test data from inspections are a valuable source of information for new owners, if the building changes ownership.

For domestic properties, a condition report can be prepared for a new owner or homebuyer. This takes the form of a visual inspection.

Efficient commissioning and reputable certification of a building's systems, coupled with planned and effective maintenance, will contribute to greater efficiency throughout the life cycle of a building.

8.3 Data management and confidentiality

Data storage

REVISED

Physical storage

In the past, physical data sources such as paper documents and drawings were the predominant method used to store and communicate information. Digital data storage and communication is increasingly utilised.

Exam tip

To understand the sort of data that can be extracted from Gantt charts and CPA documents, create your own examples for a simple activity such as making a sandwich.

Typical mistake

The terms 'commissioning', 'testing' and 'inspecting' can easily be misapplied in relation to BSE. Remember:
+ commissioning comes first, when a new system is set up ready for operation
+ testing ensures the system is operating to specified standards
+ inspection is undertaken periodically throughout the life cycle of the system to ensure the specified standards are maintained.

Now test yourself

6 What information is provided by product data and manufacturers' specifications?

7 What are the benefits of the CDE in a BIM system?

TESTED

There can be a crossover between physical and digital information sources if paper documents are scanned and converted to a digital medium for easier transmission to others in different locations.

Physical storage of confidential data

Data sources can contain sensitive information which must be kept confidential. For example, data for a construction project commissioned by the government or a bank may contain sensitive details which could threaten security if accessed by unauthorised parties.

Sensitive data stored in a physical format must be locked away and protected in secure facilities.

Virtual storage

Virtual storage is the storage of data in a digital format, for example:
+ on a computer hard drive
+ on a portable flash drive
+ in a location that is remote from the user.

Remote storage of data is often referred to as cloud storage or 'in the cloud'. Data is transmitted digitally through a network, which may include the use of the internet. Storage facilities can use digital management systems (called servers).

Data storage companies must provide safe and secure storage and have facilities in place to back up data to prevent loss.

> **Revision activity**
>
> Write a short report (around 100 words) describing the difference between physical data storage and virtual data storage.

Confidentiality

REVISED

To ensure data storage is not compromised, company procedures must be rigorously designed and carefully adhered to.

Common threats

+ The term 'hacking' has become increasingly familiar, referring to theft or manipulation of critical data, usually with malicious intent.
+ Cyberattacks are the work of individuals or organisations who steal data or install damaging software known as malware. Malware can behave like a virus, in that it can easily be passed from computer to computer, or system to system, so that the 'infection' spreads.
+ A Trojan horse or Trojan is a type of malware that is often disguised as legitimate software, misleading the users as to its true intent.

Data storage requirements

To maintain confidentiality and security of data, especially in the digital domain, sophisticated systems are used to protect information sources. As well as digital security systems, provision to keep digital devices in a physically secure location may be necessary.

Security measures and the software they use are constantly in need of updating and modifying in order to meet the growing challenge of attacks. These may include:
+ encryption of data – this uses complex mathematical methods to prevent unauthorised access
+ virus protection software – this can be of increasing complexity and must be constantly updated
+ firmware within computer systems – this must be protected and updated to prevent unauthorised interference in computer functions.

Recovery of lost or stolen data is an important activity that uses specialist techniques to defeat the attacks of hackers and data thieves.

> **Malware** Software designed to disrupt, damage or allow unauthorised access to a computer system
>
> **Encryption** Process of converting data or information into a code to prevent unauthorised access
>
> **Firmware** Computer program that makes a device work as the manufacturer intended

Legal requirements

Data protection in the UK is governed by the UK General Data Protection Regulation (UK GDPR), which came into effect on 1 January 2021. It should be considered alongside the Data Protection Act 2018.

The aim of this legislation is to ensure that personal data is gathered legally, and that those who collect and hold data protect it from misuse and exploitation.

Now test yourself

TESTED

8 Outline the methods used for virtual storage of data and information.

9 Why is malware sometimes referred to as a Trojan?

Typical mistake

It is incorrect to state that a company or individual *may* have to follow UK GDPR. All companies that sell goods or services *must* comply with this legislation.

Exam checklist

In this content area, you learned about the following:
+ data
+ sources of information
+ data management and confidentiality.

Exam-style questions

Short-answer questions

1 Give a definition of metadata. [1]

2 Explain the purpose of generalisation in analysing data. [1]

3 Give a definition of malware. [1]

4 Give a definition of a client specification. [2]

5 List **two** steps needed to make data fit for purpose. [2]

6 List **two** benefits of using a BIM system to store and process data. [2]

7 How can data be used to calculate outcomes? [2]

8 Explain what is meant by commissioning and certification data. [2]

9 List **two** measures a construction company could take to protect its digital data storage facilities from common threats. [2]

10 Explain why interoperability between digital systems is important. [4]

11 A client is due to accept handover of their new office building. What data and information will the client require from the project developer? [4]

12 What are the characteristics of a Trojan in relation to digital storage and processing systems? [4]

Extended-response questions

13 Your employer is looking for ways to improve the company's market competitiveness. Explain what areas of business activity could be analysed using accurate and current data. [9]

14 Describe the purpose and benefits of using data for post-occupancy evaluation (POE) to the contractor and the building client/owner. [9]

15 Explain why a Gantt chart or critical path analysis can be described as a data source when planning and monitoring the productivity and progress of a construction project. State why these documents are essential to efficient management of both small and large construction projects. [12]

9 Relationship management in construction

9.1 Stakeholders

Any group or individual that has an invested interest in the long-term success of an organisation is referred to as a stakeholder. Stakeholders can affect and be affected by the achievement of an organisation's business objectives, therefore it is important that they can work together.

Stakeholders can be internal or external.

Internal stakeholders:
+ support or have concern for an organisation
+ benefit from their direct relationship with the organisation, for example employers and employees.

External stakeholders:
+ do not have a direct relationship with an organisation
+ are not employed by the organisation
+ can still have an indirect effect on the organisation.

A building control officer enforcing building regulations is an example of an external stakeholder, and will determine the standards that a contractor has to adhere to.

Whenever possible, internal and external stakeholders should be selected to suit specific construction projects because of their skills, knowledge and experience working on similar jobs.

Now test yourself

1 Explain the term 'external stakeholder'.

TESTED

Typical mistake

The definition of a stakeholder is not an individual with a financial interest in an organisation. Rather, it is any group or individual that has an invested interest in the long-term success of an organisation.

Exam tip

The examiner will be looking for quality not quantity in your answers, so try to keep to the point and avoid writing more than is necessary. For example, if you are asked a question about a particular stakeholder, then structure your response around that stakeholder and avoid writing all you can about a range of stakeholders, because you will waste exam time and will not get any additional marks.

Revision activity

Record yourself reading this section and play it back to yourself several times. Alternatively, ask someone else to read it to you. Studies show that some students learn more by listening than by reading.

9.2 Roles, expectations and interrelationships

The roles, expectations and interrelationships of all stakeholders throughout a construction project include:
+ hierarchy of project management
+ promoting good relationships across the project
+ cost-control measures
+ time-management methods
+ handover processes
+ Corporate Social Responsibilities (CSR)
+ Section 106
+ follow-up and review.

Section 106 A legal agreement between a local planning authority and a landowner/ developer granting planning permission with obligations to support its aims of improving local services, e.g. creating and equipping a play area on a housing site

Check your understanding and progress at **www.hoddereducation.co.uk/myrevisionnotes**

9.3 The importance of collaborative working to project delivery and reporting

The importance of a collaborative approach

REVISED

A collaborative approach to project delivery and reporting is essential to ensure work is completed on schedule, within budget and to the client's specification.

Reporting any foreseeable problems before they occur, and working with the project team to resolve them, is essential for meeting project aims and objectives.

Collaborative working also ensures projects are completed to minimum industry standards and building regulations.

Project teams and other stakeholders that communicate effectively at every stage of construction work are more likely to provide a safe working environment for employees and others that may be affected by work activities.

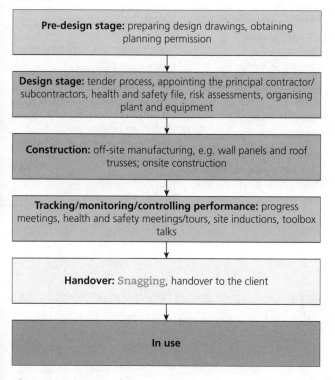

Pre-design stage: preparing design drawings, obtaining planning permission

Design stage: tender process, appointing the principal contractor/subcontractors, health and safety file, risk assessments, organising plant and equipment

Construction: off-site manufacturing, e.g. wall panels and roof trusses; onsite construction

Tracking/monitoring/controlling performance: progress meetings, health and safety meetings/tours, site inductions, toolbox talks

Handover: Snagging, handover to the client

In use

> **Snagging** Checking for minor faults in building work or materials

Figure 9.1 Stages of construction work

Collaborative working in practice

REVISED

There are various ways of collaborating on a building project, from simple face-to-face meetings to the use of information technology. These are considered in detail in section 9.10.

Building Information Modelling (BIM)

BIM is often used to collaborate at various stages of a construction project, by identifying design problems and communicating efficiently with the project team.

Workflow software packages

The construction management team can also use workflow software packages to plan and organise work. Information can then be shared easily with others and worked on as a living document to bring a project in on schedule.

Face-to-face meetings

Regular face-to-face meetings provide a valuable opportunity to discuss how a project is progressing and any issues that need addressing.

> **Now test yourself** TESTED
>
> 2   Why is a collaborative approach to project delivery and reporting important?
> 3   What digital technology is used to share construction documentation and provide a platform for collaboration?

9.4 Customer service principles

It is important for contractors to maintain good working relationships with their clients:

+ Contractors need to create a good first impression. This can be achieved through a portfolio of successfully completed contracts and positive testimonials from previous clients.
+ Contractors should be able to demonstrate good product knowledge to their clients because clients are unlikely to be experts and will be looking for guidance.
+ When a contractor works with a client for the first time, it is essential to establish a good level of trust. This can start during planning by maintaining good lines of communication.
+ If a contractor manages the client's project efficiently by meeting agreed timescales and working with honesty and integrity, then their professional relationship will grow.

Customer service principles are summarised in Figure 9.2.

Testimonials Statements of recommendation produced by satisfied customers or clients that confirm the quality of a product or service

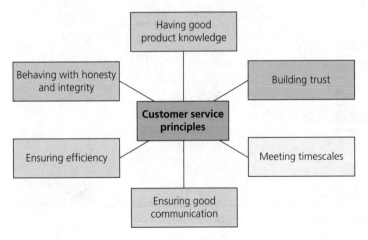

Figure 9.2 Customer service principles

The benefits of good customer service

Dealing with clients is just as important as managing a project itself, especially if they have a complaint. The way a contractor responds to negative feedback is important; they need to demonstrate that they are continuously looking for ways to improve to:
+ keep clients satisfied
+ maintain a good reputation
+ receive repeat business
+ keep their employees satisfied.

Now test yourself
TESTED ○

4 What term is used to describe statements of recommendation produced by satisfied customers or clients that confirm the quality of a product or service?

9.5 The importance of team work to team and project performance

When employees collaborate towards common goals without friction or conflict, they create a healthy work environment that improves staff morale. Employees who feel part of a team and are offered the chance to be creative and learn new skills develop positive 'can-do' attitudes that contribute to the long-term success of an organisation.

Teams perform best when:
+ individual members are motivated by satisfaction with their roles
+ everyone has accountability for the part they play
+ team members communicate openly without fear of reprisal
+ members work towards common goals.

Individuals should be comfortable both asking their peers for help when needed and providing feedback to the team when opportunities arise. This will create an environment that improves trust between team members. Conversely, conflict, tension, low engagement and lack of trust in a team will have a negative impact on project performance.

Exam tip

Try to use key words and correct terminology in your answers to demonstrate your underpinning knowledge. For example, if you are writing about examples of collaborative working practices to share construction information, you might want to mention BIM.

Now test yourself
TESTED ○

5 What impact does team work have on project performance?

Revision activity

Practise answering five written questions against the clock to prepare for your exam. Your exam will be timed, so it is important to respond with the correct answers as quickly as possible.

9.6 Team dynamics

Team dynamics refers to psychological processes and behaviours occurring in a team that influence its direction and performance.

People's personalities and behaviour are often difficult to control and unpredictable, and they can have either a positive or negative influence on team dynamics. With the right strategies in place to address any issues early on, a good team leader can:

+ manage relationships between team members
+ keep lines of communication open
+ ensure active participation and co-operation.

When all members of a team work collectively and listen to and support each other, they are more likely to resolve challenges and achieve objectives. Signs of positive team dynamics are when members work together without conflict and trust the expertise, knowledge and abilities of individuals within the team.

Businesses rely on positive team dynamics to generate new ideas and improve performance. It is therefore important to identify poor team dynamics as soon as possible. Poor team dynamics can:

+ cause individuals to feel vulnerable
+ make individuals less reliable
+ reduce individuals' ability to adapt to changing situations that may arise during building projects.

The qualities and characteristics of positive and negative team dynamics are summarised in Figure 9.3.

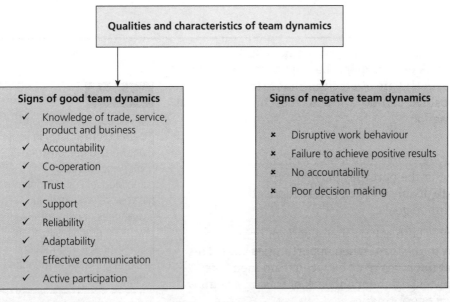

Figure 9.3 Qualities and characteristics of positive and negative team dynamics

Now test yourself — TESTED ◯

6 Why are positive team dynamics important within construction businesses?

Exam tip

Team dynamics in the construction industry are no different from those found in other teams. Try relating the principles of team dynamics to a team you are already part of (for example a sports team) to help you answer questions in your exam.

9.7 Equality, diversity and representation

Employing a talented, diverse representation of society allows different views and ways of thinking to be shared within an organisation.

Employers usually have a written equality and diversity policy, which should be read and signed by all employees as part of their induction.

Discrimination is the unfair treatment of someone because of their characteristics. For example, an employer could openly discriminate against certain groups or individuals while recruiting new employees or promoting existing employees within their organisation.

Businesses have legal duties regarding equality and diversity, as outlined in Table 9.1.

Table 9.1 Equality and diversity legislation

Legislation	Details
Equality Act 2010	This is the main legislation that protects people from discrimination at work and in wider society. Under the act, the following characteristics are protected: + age + disability + gender reassignment + marriage/civil partnership + pregnancy and maternity + race + religion/belief + sex + sexual orientation.
Employment Rights Act 1996	This act legally protects the personal rights of employees and workers. It covers areas such as: + contracts of employment + protection of wages + zero-hours workers + Sunday working + flexible working + rest breaks + study and training + unfair dismissal + maternity/parental leave + redundancy payments.
Employment Act 2008	This act covers: + the procedure for resolution of employment disputes between employers and staff + compensation for financial loss in cases of unlawful underpayment or non-payment of staff + the enforcement of minimum wages + the enforcement of laws under the Employment Agencies Act 1973 + the right of trade unions to expel or exclude members on the grounds of membership of a political party.
Human Rights Act 1998	This act sets out the fundamental rights and freedoms that everyone in the UK is entitled to, including: + freedom from torture or inhuman or degrading treatment + freedom from slavery + the right to a fair trial in a court of law + the right to respect for private and family life, home and correspondence + freedom of thought, conscience and religion + freedom of expression + freedom of peaceful assembly and association + the right to an effective remedy in a national court + freedom from discrimination + the right to education.

Exam tip

Show the examiner you have a broad understanding of the different protected characteristics and how a diverse representation in a business can have a positive influence.

Typical mistake

Equality is not just about equal rights for men and women. It is also about age, disability, gender reassignment, marriage/civil partnership, pregnancy and maternity, race, religion/belief, sex and sexual orientation.

Now test yourself   TESTED ◯

7   Define discrimination.
8   Which legislation protects the personal right of employees and workers to maternity/parental leave?

Revision activity

Research and explain the term 'positive discrimination' in the workplace.

9.8 Negotiation techniques

How negotiation techniques are used within the construction industry

REVISED

Negotiation skills are important in the construction industry. For example:

+ When buyers acquire land to build new housing, negotiations between buyers and landowners must be managed carefully. A poorly negotiated deal may reduce profit margins when completed properties are sold.
+ Once land has been acquired, planning permission may be needed before building work can start. This involves the project planning team working closely with the local planning department to develop and negotiate initial concepts into a mutually agreeable proposal.

Strategic negotiations will also be used at other stages of planning and construction, for example:

+ throughout the tendering process before awarding contracts
+ when negotiating change orders to the agreed contract between the contractor and client
+ when agreeing time extensions
+ when resolving disputes.

Different negotiation techniques

REVISED

Different negotiation techniques are described in Table 9.2.

Table 9.2 Different negotiation techniques

Technique	Description
Distributive negotiation	This is used to haggle over a common single interest at stake, known as a fixed sum. A fixed sum is best described as a pie that parties are battling over for a bigger slice, with exchange offers back and forth.
Win–lose approach	This is commonly used to settle disputes between two parties. However, an agreement is more difficult to reach because one side must compromise for the other to experience a positive outcome.
Lose–lose approach	Sometimes during business negotiations, all concerned parties end up worse off and not achieving their desired result. In these situations, all participants should try to minimise their losses as far as possible, while trying to maintain their relationship.
Compromise approach	When negotiators are unable to reach a mutual agreement and have nothing more to negotiate with, they often make **concessions** to meet the needs of the other party or to get something else they want.
Integrative negotiation (integrative bargaining)	This takes place between parties with common interests in order to collaborate in finding a mutually beneficial solution.
Win–win approach	Negotiators with shared interests work together to find resolutions they are both satisfied with, to avoid disagreements and maintain strong relationships.

> **Exam tip**
>
> Make sure you know the difference between command words, otherwise you may lose marks, for example by describing when the question asks you to explain. Note – there are some examples of different descriptions in Table 9.2.

> **Revision activity**
>
> You might find it helpful to try to explain negotiation techniques, and how these are used in the construction industry, to somebody else. Ask the other person for positive feedback and suggested areas for improvement.

Concession Something granted in response to demand

> **Now test yourself**
>
> 9 Which negotiation technique is used to haggle over a common single interest at stake, known as a fixed sum?
>
> TESTED

9.9 Conflict-management techniques

Common reasons for conflict

REVISED

Disputes can happen for many reasons, such as:
+ ambiguous contract terms
+ breaches of contract conditions
+ late supply of building materials, resources or equipment
+ breaches of site rules
+ programme delays.

> **Ambiguous** Unclear and difficult to understand

Conflict-management techniques

REVISED

One way to avoid or control conflict before it escalates is to identify issues or changes to be made using digital methods such as BIM. If a dispute does occur, each side will compete for their best interests using one of the following techniques:
+ compromise
+ problem solving
+ competing
+ forcing.

Alternatively, the following conflict-management techniques can be used:
+ informal discussions
+ mediation
+ conciliation
+ arbitration.

> **Exam tip**
>
> You will not be expected to remember all the common reasons for conflict. However, the examiner will expect to see examples of conflict-management techniques that can be used to resolve issues.

> **Typical mistake**
>
> Conflicts are not always resolved amicably. Disputes may have to be resolved using mediation, conciliation or arbitration.

> **Now test yourself** TESTED
>
> 10 Why should conflict be avoided whenever possible?

9.10 Methods and styles of communication

Methods of communication

REVISED

Communication can be either verbal or non-verbal. Inaccurate, ambiguous, incomplete or confusing information can lead to delays, missed deadlines and additional project costs. Negative communication within a business often creates passive-aggressive behaviour in individuals and a toxic culture, impacting on the business's performance and an increase in staff turnover.

> **Passive-aggressive** Indirect behaviour expressing negative feelings and hostility

Methods of communication used in the construction industry include:
+ face-to-face
+ email
+ letter
+ telephone (land line)
+ mobile phone
+ drawn information.

The strengths and weaknesses of different communication methods are discussed in Table 9.3.

Table 9.3 Strengths and weaknesses of different communication methods

Communication method	Strengths	Weaknesses
Verbal communication	+ Quick and simple + No reliance on technology, e.g. internet connection + Allows two-way conversation + Personal + Direct + Allows the recipient to confirm they have understood the information (give feedback) + Can be combined with non-verbal communication, e.g. raised voice, gestures or body language + Allows the use of both **open questioning** and **closed questioning**	+ No written record + Information can be forgotten by the recipient + Information can be misunderstood if the sender has a strong accent or regional dialect + Language barriers may exist between workers who speak different languages + Information may be misheard due to background noise + Recipient may be hard of hearing + Message could be ambiguous
Written communication	+ Can be referred back to + There is a permanent record of the communication + The same information can be distributed easily without diluting it + Sender does not have to meet the recipient to pass on the information + Can be used to communicate if the recipient is hard of hearing	+ May be no opportunity to feed back quickly + Cannot know immediately if information has been received and understood + Impersonal + Indirect + Recipient may have dyslexia, limited vision or other reading difficulties + May be language barriers + Takes time to write + Slow to distribute + Message could be ambiguous
Visualisations (graphics)	+ Message can be repeated to a wide audience + No language barrier + Message is clear and consistent + Quick to interpret a simple message + There is a written record + Eye-catching + Jargon-free + Symbols and pictograms are often standardised in the UK construction industry	+ Recipient may have a limited vision + No confirmation of acknowledgement + Easily dismissed + Recipient may need training to understand symbols, pictures or pictograms + Can get lost, removed or defaced if on display + Recipient cannot ask questions + No immediate feedback + Only simple messages can be conveyed

Open questioning Asking questions that allow someone to give a free-form answer

Closed questioning Asking questions that require a response from a limited set of answers such as 'yes' or 'no'

Exam tip

Do not waste your time trying to remember all the strengths and weaknesses for each method of communication; one example of each will be sufficient.

Styles of communication

REVISED

Communication styles can be formal or informal:

+ Formal communication is used by organisations to pass information through prescribed official channels, following an organisational structure.
+ Informal communication is much quicker because there are no rules restricting which direction or lines of communication must be used; therefore, it is 'free flowing'.

Typical mistake

Passing information from one person to another does not mean that it has been fully understood. Make sure you learn the various ways in which information can be misunderstood.

Now test yourself

TESTED

11 Describe 'passive-aggressive behaviour'.

9.11 Employment rights and responsibilities

Employment rights

REVISED

Every employee in the UK has legal rights and responsibilities under the Employment Rights Act 1996. Employment law controls employees' rights at work and relationships between employees, employers, trade unions and the government.

Figure 9.4 lists the different areas that come under employment rights legislation.

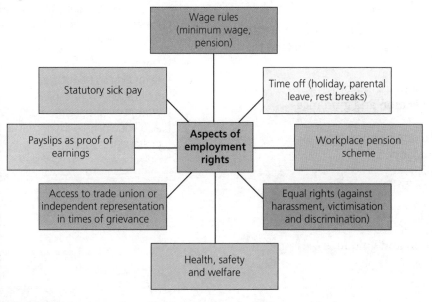

Figure 9.4 Aspects of employment rights

Exam tip

Learning the different aspects of employment rights is essential.

Typical mistake

Self-employed people do not have the same rights as employees.

119

Employment responsibilities

Employment responsibilities are summarised in Table 9.4.

Table 9.4 Employment responsibilities

Employer's responsibilities towards their employee	Employee's responsibilities towards their employer
+ Protect health, safety and welfare + Provide a contract of employment + Inform and consult when necessary + Be an inclusive employer and not discriminate + Pay minimum wage, sick pay, maternity pay, holiday pay and other entitlements + Allow staff to return to the same job after a leave of absence + Follow the Working Time Regulations 1998 + Abide by the terms and conditions of the contract of employment + Consider requests from staff for flexible working + Auto-enrol employees into a workplace pension	+ Work to the terms and conditions of the contract of employment, e.g. regarding confidentiality and reasonable behaviour + Comply with the employer's health, safety and welfare policy

Revision activity

Write all the employment rights and responsibilities on separate pieces of card. Place the cards face down and shuffle them. Remove one card and place it to one side. Turn the other cards over and try to guess which one is missing. Check your answer.

Now test yourself

12 What employment responsibilities does an employee have towards their employer?

TESTED

9.12 Ethics and ethical behaviour

Large construction businesses often develop a system of policies and practices to ensure they operate in a fair, moral and legal way; this is known as a code of ethics.

Businesses with a consistent set of values and robust moral code develop a strong ethical culture, which means they are less likely to face the consequences of unethical behaviour.

Examples of ethical behaviour in business include:
+ adherence to health and safety legislation and guidance
+ delivering equality and diversity
+ fulfilling the terms of building and employment contracts
+ honesty, integrity, commitment and loyalty
+ corporate social responsibility
+ environmental protection
+ accurate business statements.

Examples of unethical behaviour include:
+ ignoring health and safety legislation and guidance
+ discrimination
+ failure to honour commitments
+ corruption
+ harassment, for example sexual harassment
+ disregard for corporate social responsibility
+ defamation of a competitor.

Ethics Moral values that govern a person's behaviour towards others

Defamation Act of damaging someone's good reputation through a false written or verbal statement, also known as libel (written) or slander (spoken)

Now test yourself

TESTED

13 What are the consequences of unethical behaviour in business?

Exam tip

Make sure you understand the link between ethics and employment law.

Check your understanding and progress at **www.hoddereducation.co.uk/myrevisionnotes**

9.13 Sources of information

How sources of information contribute to knowledge sharing/stakeholder experience

REVISED

Networking is a low-cost process used by many organisations to:

+ introduce like-minded people
+ share ideas and knowledge
+ form long-lasting business relationships.

Construction businesses can:

+ network on construction sites
+ attend trade events or exhibitions
+ join networking groups
+ use social media.

> **Networking** Activity where businesses and people with a common interest meet to share information and develop contacts

Web-based networks

REVISED

The internet is used by construction businesses to:

+ market their product or service
+ raise their profile
+ develop people's interest for future sales opportunities.

A professionally made and managed website is essential as a platform to advertise and promote a business.

Businesses that use social media often provide customers with the opportunity to write reviews, as a way of giving unfiltered honest feedback, although it may not always be positive or constructive.

> **Exam tip**
>
> Remember that there can be many barriers to networking. Revisit section 9.10 to remind yourself of the advantages and disadvantages of different forms of communication.

> **Typical mistake**
>
> The internet does not always provide trustworthy sources of accurate information. It is important to check the validity and reliability of the sources you use.

> **Exam checklist**
>
> In this content area, you learned about the following:
>
> + stakeholders
> + roles, expectations and interrelationships
> + the importance of collaborative working to project delivery and reporting
> + customer service principles
> + the importance of team work to team and project performance
> + team dynamics
> + equality, diversity and representation
> + negotiation techniques
> + conflict-management techniques
> + methods and styles of communication
> + employment rights and responsibilities
> + ethics and ethical behaviour
> + sources of information.

9 Relationship management in construction

121

Short-answer questions

1 An employee has been unfairly dismissed in the workplace. State the legislation that protects the personal rights of the employee.   [1]

2 Which regulations state that employers have a legal duty to ensure that information is easy to understand to protect the health, safety and welfare of people from work activities?   [1]

3 Explain the term 'equality' in the context of employment.   [1]

4 A dispute has occurred between a contractor and their supplier because of some damaged goods that have been supplied. Neither party wants to go to court because of the financial expense, but both parties need to resolve the situation. Determine an appropriate conflict-management technique that could be used.   [2]

5 Identify the process to follow if discrimination was witnessed in the workplace by an employee towards their colleague.   [2]

6 The law states that employers must give their employees payslips. Explain the purpose of a payslip.   [2]

7 Explain the benefits of collaborative working on a construction project.   [2]

8 Describe integrative negotiation (integrative bargaining) and explain when it might be used.   [2]

9 You have been asked by your line manager to assess the dynamics of the construction team on site. How would you determine team dynamics?   [2]

10 Explain the importance of team work to team and project performance.   [3]

11 Large businesses often develop a system of policies and practices known as a 'code of ethics'. What is the purpose of a code of ethics?   [3]

12 Describe an external stakeholder and explain their influence on the success or failure of a construction project.   [3]

13 List **six** protected characteristics under the Equality Act 2010.   [3]

14 List **four** signs of negative team dynamics.   [4]

15 Explain the importance of a contractor establishing a good level of trust with their customers.   [4]

16 Explain why it is important for businesses to employ a diverse representation of society.   [4]

17 Give **four** strengths of visualisations (graphics) as a form of communication in the workplace.   [4]

Extended-response question

18 The principal contractor on a construction project has employed some non-English-speaking workers and others that speak English as a second language. Determine how the principal contractor should effectively communicate with these employees.   [9]

10 Digital technology in construction

The use of digital technology has led to advances in management and production methods in the construction industry, as shown in Figure 10.1.

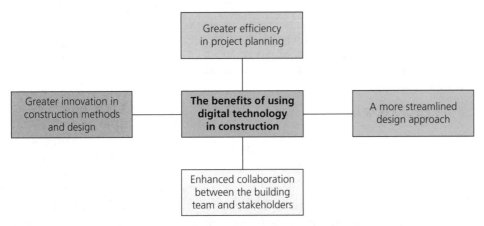

Figure 10.1 The benefits of using digital technology in construction

10.1 Internet of things

The term 'internet of things' (IoT) refers to the system of applying unique digital identifiers to physical objects, such as buildings, which can be connected to a digital network.

Objects that are connected to each other through a digital network can be referred to as 'smart' objects when the processing of information and data happens without human intervention.

Data flow between the connected objects allows communication to take place directly from machine to machine (M2M.) This interactive activity can achieve specific goals, including:

+ monitoring and optimising systems performance
+ detecting systems inefficiencies
+ alerting remote operators to potential problems
+ diagnosing possible causes of problems.

The use of technology to capture data

REVISED

The IoT can be applied to the design of buildings and infrastructure elements that incorporate interconnected sensors and monitoring equipment. This allows data analysis that can translate into:

+ reduction in energy use
+ improvements in manufacturing efficiency
+ improvements in safety
+ streamlining of materials delivery and supply-chain activity.

Smart technology

Smart technology can help building designers to identify optimal construction materials and methods for future projects, leading to the construction of buildings that:

+ rely less on highly skilled workers
+ use fewer resources
+ reduce waste
+ produce less carbon.

123

Capturing and analysing data

The interconnected nature of the IoT allows data analysis which can provide performance information on individual components and materials in a range of structures.

Manufacturers utilising the IoT could potentially identify performance issues with a view to improving and refining manufactured resources.

Artificial intelligence (AI)

Artificial intelligence (AI) describes digital systems that go beyond simply following programmed instructions. They use methods of analysis that can identify patterns and correlations in order to draw appropriate and pertinent conclusions more quickly than humans are able to.

AI could also be used during the design of a new building to make comparisons of:
+ different design features
+ different construction methods
+ material and component costings
+ systems performance factors.

> **Revision activity**
>
> AI is used in face- and fingerprint-recognition software that is often used in mobile phones. Go online and use the search term 'Fingerprint recognition in construction', then write a short report on how this digital technology is used to keep construction sites secure.

Digital technology used in completed buildings

REVISED

Completed buildings that are designed to integrate with the IoT can produce large quantities of useful data.

Sensor technology and data management systems embedded in a structure can record and archive data, as well as produce live data streams for analysis and evaluation. This allows a clear understanding of the:
+ performance of an occupied building (for example energy use)
+ environment that the building provides for its occupants.

Occupier comfort

To ensure a comfortable and productive working environment in buildings such as offices, smart control systems can be used to monitor and adjust:
+ relative humidity
+ air temperature
+ ventilation levels
+ dust levels
+ air pressure within individual rooms.

Power management

Demand for electrical power in an occupied building will fluctuate depending on the number of occupants active at certain times. Using intelligent systems, digital technology can:
+ gather and compare information over time to detect fluctuations in energy use
+ predict periodic peaks and troughs in demand
+ meet energy demands more efficiently and economically.

Maintenance

AI systems can compare a huge range of scenarios based on live data streams, from which warning signs of equipment failure can be identified. This allows predictive maintenance to be a reality.

The development of AI systems that can detect potential faults and anomalies in equipment means that maintenance interventions can be made before that equipment fails.

> **Revision activity**
>
> Write a summary of ways in which AI can be used to manage energy use in large buildings.

Check your understanding and progress at **www.hoddereducation.co.uk/myrevisionnotes**

Digital technology used during construction of buildings

Design

Computer-aided design (CAD) has become a standard digital design tool when developing the concept for a construction project. Digital systems can be used to:

+ develop the concept
+ speed up the design process
+ test and refine the design concept.

> **Exam tip**
>
> To better understand how effective CAD can be in creating and refining the design concept of a building, try a free CAD program such as SketchUp for yourself.

> **Typical mistake**
>
> When answering questions on the use of digital technology in the design of buildings, keep in mind that the value of human input is not diminished. Producing the project concept still needs imagination and creativity – something digital technology and AI are not capable of … yet.

Digital tools are able to generate accurate materials and components lists from 3D models or 2D drawings. These can be linked digitally to materials costs databases in order to produce an up-to-date costing for a project.

Speeding up the process of designing and costing a project, coupled with the ability to update or amend the project details quickly, streamlines the design development and improves productivity.

On site

Examples of smart equipment used (or being developed for use) on site are listed in Table 10.1.

Table 10.1 Examples of smart equipment for use on site

Type	Description
Surveying tools	These use satellite positioning data for setting out buildings. They can link with remotely stored project data through the IoT, which can speed up site preparation considerably.
Mechanical excavators	These can be equipped with satellite positioning equipment and digital depth controls to avoid damaging buried services.
Earth-moving and excavation machinery	This works with pinpoint accuracy in levelling the ground and creating gradients and slopes when contouring the terrain.
Tracking equipment	This is installed in machinery to assist in planning the deployment of the right type of equipment to the required site area.
Systems that create a virtual boundary	A virtual boundary (or **geofence**) ensures that machinery only works in specified areas.

> **Exam tip**
>
> If you are asked about smart equipment, you may need to mention that some types described here are not yet widely used or are under development.

> **Geofence** Virtual perimeter for a literal geographic area

Materials and resources

Delivery of materials and components to site must be reliable and consistent in order to avoid delays and increased costs.

Just in time (JIT)

Storage of materials on site is often limited due to congestion and a lack of suitable storage space. A strategy developed in the car manufacturing industry known as 'just in time' (JIT) can be used in construction activities where storage facilities or access are limited.

Use of JIT can:
+ improve efficiency
+ reduce waste
+ eliminate the need for extensive onsite storage facilities.

A JIT strategy requires delivery of the right materials, in the right order, in the right amount, at the right time.

Digital technology can be used to track and monitor materials and components during manufacture, transport and delivery.

Now test yourself   **TESTED**

1 What does M2M mean?
2 List **four** potential benefits of creating an M2M system connecting buildings.
3 List **two** ways in which AI can be used in construction.
4 How can digital technology be used to enhance comfort levels for a building's occupants?
5 How can AI be used to predict equipment failure in a building?
6 Identify smart equipment and the way it can be used on site.

> **Revision activity**
>
> Explain the benefits of a JIT system and how it works.

10.2 Digital engineering techniques

Digital engineering uses a digital skillset to create and edit data as part of a design process. Engineering principles are applied in a virtual environment, allowing building designers to explore a range of possible design options and develop innovative solutions to design problems.

There are three main techniques applied in the field of digital engineering:
+ simulation
+ animation
+ modelling.

> **Exam tip**
>
> A scenario may be used as the basis for questions about simulation, animation or modelling. Take a moment to picture what the scenario describes before answering.

Simulation   REVISED

Simulation uses digital methods to mimic the behaviour of real systems and processes. It can answer important questions about a proposed construction project, such as:
+ Will the building design be energy efficient?
+ Which construction methods are best suited to the design concept?
+ What is a realistic timescale for the construction of the building?

The sequencing of project stages can be simulated, so that construction processes follow a smooth path without disruptive clashes between operations. Any necessary changes can be made before work commences on site, avoiding errors and expensive delays.

Simulation can be used for structural analysis. Performance data harvested from past projects can be used to model the behaviour of materials and components in the planned project to ensure that the proposed design can carry the stresses and loadings that will be imposed on it.

When retrofit or restoration work is planned, simulation can be used to ensure that the existing structure can accommodate new materials or components without overloading or stressing structural elements.

> **Restoration** Process of returning a building to its original condition

Animation

REVISED ○

Modelling a structure in 3D using digital technology can be used for refining conceptual details of a project at the design stage.

While a 3D digital model can deliver an impressive view of a project, the viewer's experience can be greatly enhanced if that model is animated. A lifelike view of a building can be created, often allowing a walk-through of the structure to show room areas and features.

Modelling

REVISED ○

Digital (or computer) modelling can be used throughout the life cycle of buildings. For example, sophisticated digital modelling is used to survey existing buildings in order to collect data that can be used for:
+ assessing maintenance and repair requirements
+ planning restoration projects
+ carrying out alterations and extensions
+ making decisions on demolition methods at the end of a building's useful life.

Use of laser scanners

Laser scanners can be used to:
+ create 3D imagery of a building's complex geometry
+ survey surface areas that are not easily accessible.

From its setup position, a laser scanner can digitally record precise measurements of densely grouped points at rapid speed, often referred to as a 'point cloud survey'. The point cloud data can be integrated into BIM software or CAD systems to digitally create detailed 3D models of a structure or features of the built environment for a range of uses.

Laser scanners can be mounted on drones to capture the exact contours of landscapes, road layouts, railway routes and even entire towns. This is often referred to as geo-surveying.

Use of scanned data in virtual reality (VR)

Scanned data can be utilised in immersive technologies, such as virtual reality (VR).

A VR user is visually completely shut off from the outside world by wearing a head-mounted display (HMD). Whatever input the user sees through the HMD becomes their 'reality', allowing them to experience a digitally generated scene as if they were part of it.

The data in digital models can also be used to create static illustrations or artistic impressions of a project for use in conventional non-digital documentation, such as a client brochure or report.

> **Revision activity**
>
> List the ways in which laser scanning can be used for modelling structures or other features of the built environment.

> **Exam tip**
>
> The benefits of using digital engineering techniques could be summed up as:
> + efficiency
> + accuracy
> + speed.

> **Immersive** Generating a three-dimensional image which appears to surround the viewer

> **Typical mistake**
>
> Other types of immersive technology are often mistakenly grouped under the title 'virtual reality'. Augmented reality (AR) and mixed reality (MR) are actually different systems that can be applied in various construction-related contexts.

> **Exam tip**
>
> There are lots of abbreviations used in digital technology, such as VR. When you come across one, write it down with its meaning as a memory aid.

Now test yourself | TESTED ○

7 How can simulation be used to ensure the construction process follows a smooth path?
8 In what ways can an animated walk-through of a building benefit the design process?
9 List **two** ways in which immersive technology can be used to benefit the client.

10.3 Adapting technologies used in other industries

Beneficial uses of technology

REVISED ●

Table 10.2 lists some of the benefits of using digital technology in construction activities, especially in relation to planning and organising a project and monitoring the progress of the construction phase.

Table 10.2 Benefits of digital technology in construction activities

Benefit	Application
Accuracy	Digital technology allows processing of large quantities of data used in: + measurement + costings + projections + evaluations. Information can be cross-referenced and linked in complex ways to produce rich and valuable assessment data. This can inform reliable decision making and forecasting during the: + planning of a project + monitoring of progress during the construction phase.
Accessibility	Digital data can be stored and retrieved quickly and easily. Data can be accessed by anyone who is authorised to use relevant networking systems, regardless of their location. This encourages collaboration and allows data to be instantly updated or modified.
Efficiency	Digital networks can be permanently open to authorised users working in local teams or global partnerships across different time zones. Effective work patterns can be created and refined to respond to current needs in maintaining efficiency when working: + in teams + in isolation + simultaneously + sequentially.
Risk reduction	Digital simulation of construction operations can be used to: + identify areas of operational risk + analyse accident and injury data to identify common elements of risk in behaviour or work patterns + identify emerging trends in safety matters to quickly introduce safer work methods + match risk assessments and method statements more closely with actual working patterns and conditions.

> **Exam tip**
>
> A table is a useful way of displaying key points in an easy-to-understand format. It may be helpful to create your own tables to memorise key points for different topics.

Robotics

REVISED ●

Industries such as car manufacturing use robotics with digital control systems to assemble components repetitively.

There are some repetitive activities in construction where industrial robots are already being used or are being developed. An example is off-site construction, where parts or sections of a building are manufactured away from the site location and then transported to site for assembly. Off-site construction is often referred to as prefabrication and may involve modular construction methods.

> **Prefabrication**
> Construction of buildings and their components at a location other than the building site

The benefits of using robots in a factory setting include:
+ consistent quality standards
+ regulated work rate allowing materials delivery to be reliably scheduled
+ reduced waste through accuracy and efficiency
+ enhanced safety standards as human involvement is reduced.

Other types of robot that could be used in the construction industry include:
+ inspection robots that can reach areas that are difficult or dangerous to access
+ maintenance robots that can clean and repair buildings safely
+ 3D-printing robots that 'print' components or whole buildings
+ demolition robots that safely dismantle unstable or fragile structures.

CAD/CAM

REVISED

CAD/CAM refers to software that is a combination of two processes:
+ Computer-aided design (CAD) is a digital tool used to create 2D and 3D representations and drawings during the design process, which allows simulation, testing and refining of ideas to arrive at a satisfactory finished design.
+ Computer-aided manufacturing (CAM) uses digital geometric data to control manufacturing machinery, such as computer numerical control (CNC) machines, which have motorised tool positioning and manoeuvring capabilities controlled by pre-programmed computers.

> **Geometric** Consisting of defined angles, patterns and shapes

Exam tip

CAM machines are essentially a type of robot, so information in the previous section can be applied here too.

CAD, CAM and BIM

Using digital technology provides opportunities for the integration of CAD, CAM and BIM, working together to save time and avoid possible human error in interpreting drawings and written instructions. For example:
+ A 2D or 3D CAD drawing of a building consists simply of lines to represent its shape and the component parts within it.
+ If the CAD data is linked to BIM, the combinations of simple lines can be allocated grouped geometric patterns, features and dimensions.
+ This can allow automatic generation of a range of accurate views of a component, along with details of the materials it is made from, so that a CAM machine can manufacture it.

> **Revision activity**
>
> Write a summary of the benefits of integrating CAD, CAM and BIM.

Computer modelling

REVISED

The digital processes referred to above are often described under the generic label of 'computer modelling'. The range of uses for this technology is potentially vast. Digital systems used in other industries will continue to be adapted for use in construction and the built environment.

Now test yourself

TESTED

10 Select **two** of the benefits of using digital technology listed in Table 10.2 and explain how they can potentially be applied.
11 Explain **two** benefits of using robots in the manufacture of prefabricated buildings.

Exam checklist

In this content area, you learned about the following:
+ internet of things (IoT)
+ digital engineering techniques
+ adapting technologies used in other industries.

Exam-style questions

Short-answer questions

1 Give a definition of the internet of things (IoT). [1]

2 The IoT can be applied to the design of buildings to incorporate interconnected sensors and monitoring equipment. State **two** benefits of doing this. [1]

3 Explain **one** way in which data analysis using AI can be used during the design of a new building. [1]

4 List **two** ways in which digital technology can be used to manage power supply and usage in a large building. [2]

5 Discuss the benefits of the use of digital technology in construction. [3]

6 What are the **three** elements of a just-in-time (JIT) system when obtaining materials? [3]

7 What does HMD stand for in immersive technology? [1]

8 What are the advantages of working with digital data in terms of accessibility? [3]

9 List **two** benefits of using robotics in assembling factory-produced construction components. [2]

10 Besides manufacturing components, what can robotics be used for in construction? [3]

11 Describe how a CNC machine works. [3]

Extended-responses questions

12 Explain how the use of digital technology can make operations on site safer. [9]

13 A building plot has been purchased to construct a retail facility in a busy city-centre location. In what ways can digital simulation be useful to the design team? [9]

14 An iconic Victorian building must be restored. Discuss how digital technology using drones and laser scanners could be used to survey the project. [12]

11 Construction commercial/ business principles

11.1 Business structures

Limited company (PLC or Ltd)

There are several benefits of public limited companies (PLC) and limited companies (Ltd), including:

+ protection of owners' personal assets from any business debts
+ owners given complete control of their business
+ no requirement to pay National Insurance
+ more tax benefits than sole traders.

However, information about the directors, shareholders, registered office and yearly financial statements is on public record

> **Exam tip**
>
> Spend time comparing not for profit organisations/community interest companies with limited companies so that you can provide comprehensive answers about these business types.

> **Public limited company (PLC)** Similar to a private limited company, the main difference being that money can be raised for the business through investors buying shares on the stock exchange

> **Typical mistake**
>
> Limited company owners do not have any personal financial liability if things go wrong.

Small and medium-sized enterprises (SMEs)

SMEs are defined as follows:
+ small: employs on average no more than 50 people and has an annual turnover of £10.2 million or less
+ medium: employs on average no more than 250 people and has an annual turnover of £36 million or less.

Not-for-profit organisations

Not-for-profit organisations:
+ are charitable businesses that do not make a profit
+ provide a public service or social benefit
+ are run by a board of directors
+ can be a limited business
+ can have tax-exemption status
+ must submit annual tax returns.

Community interest company (CIC)

Community interest companies (CICs) aim to:
+ benefit the community or trade with a social purpose
+ make a profit to reinvest in the company or community.

Owners and investors are allowed to make a reasonable return.

> **Revision activity**
>
> Create a simple table, using each business type as the column headings. In each column, write a bullet list of the business principles.

11.2 Business objectives

Business objectives can be defined as the incremental steps a business needs to take in order to achieve its overall aims, which are often closely aligned to its business plan.

Objectives should be challenging, to give a business purpose and direction. Once they have been achieved, they should be updated to move the business one step closer to achieving its main goal.

> **Business plan** Written document that defines a business's goals and the strategies and timeframes to achieve them

131

Examples of business objectives used to measure the performance of an organisation in the construction industry are listed in Table 11.1.

Table 11.1 Business objectives used to measure the performance of an organisation

Objectives		Examples
Financial	Private organisations	+ Maximise profit + Improve cash flow + Grow revenue and earnings + Develop the business with innovative ideas, products or services + Ensure wider profit margins + Achieve market leadership
	Not-for-profit organisations	+ Cover costs + Reduce poverty + Establish reserves + Fund activities that benefit the community + Create value for money + Increase access + Reinvest in the business
Social	Private organisations	Provide employment
	Not-for-profit organisations	Provide: + housing + services + education + healthcare
Organisational culture		Align the following with business objectives: + beliefs + ethical values + behaviours
Innovation		+ Allow for the generation of ideas + Align innovation activities and goals with business objectives + Improve the quality of a service or an existing product
Quality		Meet the standards required by the following quality marks: + Building Research Establishment's Home Quality Mark (HQM) + International Organization for Standardization (ISO)
Sustainability		Embed sustainability into business objectives, e.g. + energy-efficient construction + eco-friendly use of materials + reduction of waste
Compliance		+ Ensure regulatory compliance with external rules + Build internal controls into objectives

Typical mistake

Business objectives are sometimes confused with business aims or long-term goals.

Exam tip

It is essential to learn different examples of business objectives.

Now test yourself

1 What is the purpose of business objectives?
2 What document defines a business's goals and the strategies and timeframes to achieve them?

TESTED

Revision activity

Imagine you are about to start a business. Describe your business's principles and explain why these are important to you.

11.3 Business values

Fundamental business values

REVISED

Business values play an essential role in any organisation. They can:
+ create a sense of purpose and commitment
+ improve cohesion of the workforce
+ drive an organisation forward
+ help motivate employees by building trust and security
+ help develop relationships with partners, stakeholders and customers

Cohesion State of working together in unity

132

Check your understanding and progress at **www.hoddereducation.co.uk/myrevisionnotes**

- influence sales, customer service and marketing strategies
- demonstrate the business's culture outside of the organisation, which may attract new talent.

The fundamental business values for construction organisations are described in Table 11.2.

Table 11.2 Fundamental business values for construction organisations

Business value	Why it is important
Financial stability	This means that a business has sufficient funds to: - pay overheads - repay any loans - make a profit to prepare for risk management in times of potential **economic downturn**.
Providing good customer service	This is vital to the success of all construction businesses, to ensure: - a good business reputation - repeat business - recommendations to other potential customers.
Care for life	The health, safety and welfare of all employees and others affected by work activities should be one of the core values of all businesses. Besides the legal obligations that businesses must adhere to, they should also consider the mental health and physical wellbeing of their staff, and the impact that their job roles may have on them.
Ethics and transparency	Business ethics can be described as business practices and policies when faced with arguably controversial subjects. Business transparency is about open and honest communication across all levels within an organisation and the sharing of information both internally and externally.
Code of conduct	This is essentially a set of rules written by an employer for their employees, to protect their business and its reputation. It should explain an organisation's values and principles and link them with standards of professional behaviour.
Collaborative working	Internal collaborative working enables a diverse range of employees with specific skills or traits to work together to problem solve and achieve business objectives. External collaborative working is referred to as networking.

> **Economic downturn**
> When the economy has stopped growing and is on the decline, resulting in reduced financial turnover

Now test yourself   TESTED ⬤

3 Why is a code of conduct important for a business?

> **Exam tip**
> Make sure your answers show that you understand the fundamental reasons for business values.

11.4 Principles and examples of corporate social responsibility

The basic principles of corporate social responsibility (CSR)

REVISED ⬤

Corporate social responsibility (CSR) is the commitment of an organisation to carry out its business activities in a socially and environmentally responsible way.

CSR is not a legal requirement. However, developing a CSR strategy that will have a positive impact on the community and wider society, and integrating it into an organisation's values, makes good business sense:

+ It often contributes to risk management and legal compliance.
+ It affects how stakeholders such as clients and investors view an organisation.
+ It may impact on stakeholders' decisions to work with or support an organisation.

Examples of use in the construction industry

Table 11.3 provides examples of how CSR is applied in the construction industry.

Table 11.3 CSR in the construction industry

Principle of CSR	Application in the construction industry
Construction design	+ Responsible purchasing + Career management + Use of local operatives/trades/suppliers and local sustainable materials + Sustainable initiatives
Economic and environmental considerations	+ Boosting the economy by increasing local employment and paying above the minimum living wage + Encouraging flexible working
Legal considerations	+ Equality and diversity through employment opportunities + Following employment rights and responsibilities, e.g. making sure employees are paid at least the national living wage + Working conditions for workers, e.g. health, safety and welfare
Ethical considerations	+ Supporting training and development programmes, e.g. work experience, apprenticeships and internships + Ensuring buildings are well designed to suit occupants' lifestyles + Being guided by the local community to meet its needs (e.g. providing functional community spaces, landscaping, cycle paths, community lighting for safety)
Philanthropy	+ Ensuring inclusivity (e.g. providing affordable homes) + Improving quality of life and wellbeing + Paying National Insurance contributions and offering workplace pensions to workers

Typical mistake

CSR is not a legal requirement. However, there may be legal duties that align with a business's CSR commitments.

Exam tip

Section 3.1 is closely related to this section, so be sure to revisit it to underpin your knowledge in preparation for the exam.

Now test yourself

TESTED ◯

4   According to CSR, construction businesses should build sustainably. How can this be applied in practice?

Revision activity

Research a local construction business to identify its corporate social responsibilities.

11.5 Principles of entrepreneurship and innovation

The role innovation and entrepreneurship play in the construction industry

An entrepreneur is an individual who starts up their own business, taking on financial risk in the hope of making a profit.

The development of innovative technologies, products and processes is extremely important to the future of the construction industry, as shown in Figure 11.1.

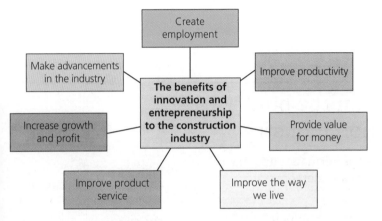

Figure 11.1 The benefits of innovation and entrepreneurship to the construction industry

Principles of innovation and entrepreneurship

Table 11.4 Principles of innovation and entrepreneurship

Principle	Explanation
Solution provider	Entrepreneurs who are willing to take a risk, and have the vision to solve problems with an innovative product or service, are likely to have a successful enterprise.
Vision	This is the business's mission – what does the business hope to achieve? A business's goals could be economic, financial, social or environmental.
Viable product/service	The aim of a business is to provide a service or product for which there is a demand or that has a unique selling point (USP).
Capital	Starting a new business often requires an investment known as **working capital**. This money could come from the owner/s, investors, shareholders or a business loan.
Growth and marketing	For a business to thrive, potential customers need to be made aware of the: + product or service being sold + people selling it + price + place where they can buy it. Therefore, some of the money invested in a business should be allocated to promoting the product or service through marketing.
Research	To be successful, a business owner must know their market well, including its strengths and weaknesses.
Priorities	Entrepreneurs with a concept for a new business must: + have a list of priorities to get their venture off the ground + become financially stable as soon as possible + grow in the future.

Working capital Sum of money remaining after all the business's debts have been covered

11.6 Measuring success

Measuring business success in the built environment and construction industry

REVISED ⬤

Successful organisations regularly evaluate their performance and make improvements, either to tackle areas of weakness or to demonstrate excellence.

The term 'benchmarking' means measuring an organisation's internal and external performance against pre-determined industry standards, competitors or completed projects. It is a powerful management tool that can increase both productivity and profits with minimum input and maximum output (benefits).

Benchmarking Measuring an organisation's internal and external performance against pre-determined industry standards, competitors or completed projects

Any business targets set by an organisation must be realistic and achievable, if the business is to accomplish its aims.

Examples of benchmarking used in the built environment and construction industry are shown in Figure 11.2.

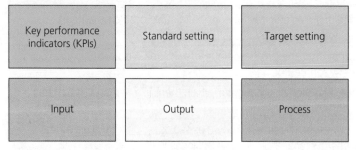

| Key performance indicators (KPIs) | Standard setting | Target setting |
| Input | Output | Process |

Figure 11.2 Examples of benchmarking

Benchmarking is a recognised system that involves identifying problem areas in a business that need improvement, for example:
+ project completion times
+ planned budgets against actual costs
+ profit margins.

Areas that can be measured in terms of a value are known as key performance indicators (KPIs). Every business selects its own KPIs, based on what is important to it.

The process of benchmarking is shown in Figure 11.3.

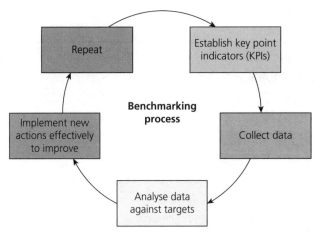

Figure 11.3 The process of benchmarking

Measuring business performance against KPIs helps to establish whether objectives have been met or whether new strategic targets need to be planned.

Gathering information for benchmarking within an organisation is much easier than trying to measure the success of a business against its competitors because of difficulties in accessing all the data needed. Internal benchmarking is often more valuable than external benchmarking because it is focused and can be tailored to meet specific business objectives.

External benchmarking is used by businesses to focus on the industry as a whole. Construction KPIs can be gathered from the results of national surveys of clients and construction professionals, and performance data based on thousands of projects across the UK.

Now test yourself   TESTED ◯

7   In the process of benchmarking, what follows analysing data against targets?

Exam tip

If you get a question on measuring success, try to mention benchmarking, and how it is used, in your response to gain the highest marks.

11.7 Project management

The principles of project management include:
+ effective planning
+ setting clear goals and objectives
+ defining roles and responsibilities
+ setting realistic milestones
+ setting constraints on cost and time
+ ensuring all objectives are **S**pecific, **M**easurable, **A**chievable, **R**ealistic and **T**ime bound (**SMART**) or following the principles of PRINCE2 (**PR**ojects **IN C**ontrolled **E**nvironments), an industry-standard process-based framework/qualification for effective project management.

Exam tip

Identify the key terms in each section of this book and make sure you understand their definitions. You will be awarded additional marks if you refer to these terms in your exam.

Typical mistake

Students lose marks because they do not understand acronyms used and are unable to answer a question in full – make sure you learn what different acronyms stand for.

Revision activity

Find an example of a Gantt chart on the internet for a simple task and use this as a template to plan your exam revision.

Now test yourself   TESTED ◯

8   What is the purpose of benchmarking?
9   What is the purpose of the acronym SMART when target setting?

11.8 Quality management

To maintain the standard or quality of work in a consistent manner, businesses use various quality management systems and techniques, and these are described in Table 11.5.

Table 11.5 Business quality management systems and techniques

Quality management system/technique	Purpose
Self-assessment	This is the proactive practice of periodically internally evaluating a business's management system to make sure it is working effectively.
	Reviewing quality management systems implemented by an organisation against a model provides an opportunity to recognise any achievements and identify areas of weakness for continuous improvement.
Internal audit	This type of quality management system is less expensive than an external audit because it is completed by the organisation itself.
	Auditors should already know the organisation and what can be achieved, therefore this can improve ownership of issues found and develop competence internally.
External audit	External auditors: ✤ are independent and should therefore not have any internal influence ✤ have a broader experience of different workplaces than internal auditors and will look at quality management systems from a different perspective.
	Recommendations from independent auditors often carry more weight.
Quality control	This system is used to ensure operational activities are performed and achieved.
	Quality control inspections are an ongoing process that enables business standards to be maintained in order to meet the organisation's and customers' expectations.
Quality improvement	See section 11.6 for the principles and purpose of target setting/benchmarking for continuous quality improvement.
ISO 9000	The International Organization for Standardization (ISO) is an independent, non-governmental organisation that develops and publishes international standards.
	ISO 9000 is a family of quality management standards designed for organisations, regardless of size, to improve the quality of products and services and help meet customer expectations.

Typical mistake

Marks will be lost if you are unable to identify a range of different quality management systems – so make sure you know these well before the exam.

Exam tip

If you specifically mention ISO 9000 when asked about quality management systems and techniques used in business, you will be awarded extra marks because of the depth of your answer.

Now test yourself

TESTED

10 What are the potential disadvantages of internal audits as part of a quality management system?

Exam checklist

In this content area, you learned about the following:
+ business structures
+ business objectives
+ business values
+ principles and examples of corporate social responsibility
+ principles of entrepreneurship and innovation
+ measuring success
+ project management
+ quality management.

Check your understanding and progress at **www.hoddereducation.co.uk/myrevisionnotes**

Exam-style questions

Short-answer questions

1 State the government department that sole traders must be registered with as soon as they start trading. [1]

2 Explain the term 'benchmarking'. [1]

3 Explain how a sole trader can protect their personal assets from any liability for losses or debts that their business may incur. [2]

4 Explain the term 'business objectives' and why they are so important to an organisation. [2]

5 Entrepreneurs often start their businesses with a business plan. Explain the term 'business plan'. [2]

6 Explain why SMART objectives must be time bound in business. [2]

7 Explain the process a sole trader would have to follow if they would prefer to use a business name that is different from their own. [2]

8 State the international standard-setting body that offers solutions to global challenges and supports innovation by providing guidelines to streamline processes and improve quality and safety. [2]

9 Describe the purpose of a not-for-profit organisation. [2]

10 State **three** methods of measuring business success. [3]

11 Explain how the principles of corporate social responsibility (CSR) can be applied in the construction industry. [3]

12 A franchise has made an end-of-year profit of £75,000 after tax. The franchisee must pay a sixth of their profits to the franchisor. On average, how much did the franchisor receive from the franchise every month? [2]

13 Give **two** disadvantages of being a sole trader. [2]

14 List the main categories of business objectives used to measure performance of an organisation in the construction industry. [4]

15 Explain the meaning of the acronym PAYE and explain the advantages of this for an employee working as a building services engineer. [4]

16 Describe the process of benchmarking used to measure an organisation's internal and external performance. [4]

17 List the benefits of innovation and entrepreneurship. [3]

Extended-response questions

18 Explain the principles of corporate social responsibility and how this is integrated into building services engineering for construction business activities. [9]

19 A trade foreman has identified that one of the PAYE electricians has not been completing the work they were expected to on each new unit (house), within an acceptable amount of time. Using SMART objectives, describe an example target that could be used to measure the performance of the same electrician on the next unit. [9]

20 A construction contractor has just received the annual results of an internal health and safety audit. The audit results are worse than the previous year due to an increased number of minor injuries involving portable power tools on sites under the control of the contractor. Analyse the effect on the business of decreasing health and safety performance and suggest a strategy that could be used to improve performance. [9]

Exam tip

The difficulty of the questions increases throughout the exam paper, so make sure you allow yourself enough time to answer the extended-response questions later in the exam.

Glossary

Access equipment Apparatus specifically designed for working safely at height

Accident book Formal document used to record details of accidents that occur in the workplace, whether to an employee or a visitor

Accredited Officially recognised as meeting professional quality standards

Aggregate Coarse mineral material such as crushed stone (gravel) used in making concrete

Ambiguous Unclear and difficult to understand

Anaerobic digestion Process by which bacteria break down organic matter

Approved Code of Practice (ACOP) Document providing advice and guidance on how to comply with health and safety law, published by the HSE

Atmospheric pressure Force exerted on the Earth's surface by the weight of air above; this varies depending on height above sea level

Benchmarking Measuring an organisation's internal and external performance against pre-determined industry standards, competitors or completed projects

Building Information Modelling (BIM) Use of digital technology to share construction documentation and provide a platform for collaboration

Building line Boundary line set by the local authority beyond which building work must not extend

Building notice Basic application form sometimes sent to a local authority planning department to inform it of the intention to complete minor building work

Building regulations Mandatory building standards in the UK

Business plan Written document that defines a business's goals and the strategies and timeframes to achieve them

CAD programs Computer-aided design software used to produce design and technical documentation

Carbon emissions Carbon dioxide released into the atmosphere, which is a cause of climate change

Centre of gravity Imaginary point where the weight of an object is concentrated

Civil engineering Profession involving the design, construction and maintenance of infrastructure that supports human activities, for example roads, bridges, airports and railways

Climate change Large-scale, long-term change in the Earth's weather patterns and average temperatures

Closed questioning Asking questions that require a response from a limited set of answers such as 'yes' or 'no'

Cohesion State of working together in unity

Commercial Relating to business; involving buying or selling

Companies House Government body that registers and stores information on all the limited companies in the UK and makes it available to the public

Concession Something granted in response to demand

Confined spaces Workplaces that may be substantially but not always entirely enclosed, where there is a foreseeable serious risk of injury because of the conditions or from hazardous substances

Continuing professional development (CPD) Process of maintaining, improving and developing knowledge and skills related to one's profession in order to demonstrate competence

Conventions Agreed, consistent standards and rules

Copper alloys Metal alloys that contain copper and one or more other metals

Corporate social responsibility (CSR) The commitment of an organisation to carry out its business activities in a socially and environmentally responsible way

Corporation Business owned by its shareholders

Corrosion Gradual deterioration of metals through chemical or electrochemical reaction with their environment

COSHH assessment Process for controlling the use of hazardous substances in the workplace

Dangerous occurrences Incidents that could have caused harm, injury or ill health

Data interoperability Ability of data systems to exchange and use information

Defamation Act of damaging someone's good reputation through a false written or verbal statement, also known as libel (written) or slander (spoken)

Design brief Document for the design of a project developed in consultation with the client/customer

Desktop survey Administrative investigation into a piece of land, completed without visiting the site

Domestic Relating to a dwelling or home

Duty holders People with legal responsibilities under health and safety law

Dynamic Characterised by frequent change or motion

Economic downturn When the economy has stopped growing and is on the decline, resulting in reduced financial turnover

Elevation View of the front, back or sides of a building

Emitters Radiators or heaters used to heat a room

Encryption Process of converting data or information into a code to prevent unauthorised access

Energy performance certificate (EPC) Document that provides an energy-efficiency rating for a building

Ethics Moral values that govern a person's behaviour towards others

Fauna Animals in a particular region

Fenestration Openings in the façade of a building, for example windows and doors

Ferrous metal Metal that contains iron

Firmware Computer program that makes a device work as the manufacturer intended

Flora Plants and trees in a particular region

Fossil fuels Finite energy sources formed by the decomposition of organic matter beneath the Earth's surface over millions of years, for example coal, gas and oil

Frontage line Front part of a building that faces a road

Full planning application Detailed formal request to a local authority planning department prior to building work taking place

Gantt chart Programme of work used to plan the sequence of building work, delivery of resources and map progress against intended start and completion dates for a construction project

Generalisation (In data processing) creating layers of summarised information from a mass of data

Geofence Virtual perimeter for a literal geographic area

Geometric Consisting of defined angles, patterns and shapes

Green roof Sustainable roof system that involves installing additional waterproof membranes and drainage mediums, onto which soil is added to allow growth of vegetation; this protects and insulates the building and reduces its environmental impact

Greenfield sites Areas of land that have not been previously developed or built on, above or below ground (sites that have previously been developed are known as brownfield sites)

Gypsum Natural mineral used in products such as cement, plaster and plasterboard

Hardcore Made from waste broken bricks, stone and/or blocks; used to provide a base

Hardness Ability of a material to resist scratching, wear and tear, and indentation

Hazard Something with the potential to cause harm

Hydroelectric power Form of renewable energy that uses the power of moving water to generate electricity

Hydrology Study of water in the earth and its relationship with the environment

Immersion heater Electrical element that sits in a body of water; when switched on, the electrical current causes it to heat up, which in turn heats up the surrounding water

Immersive Generating a three-dimensional image which appears to surround the viewer

Improvement notice Legal document issued by the HSE to an employer, instructing them to put right within a specific period of time any health and safety faults identified

Infrastructure Basic systems and services required for the proper functioning of society

International Organization for Standardization (ISO) Independent, non-governmental organisation that develops and publishes international standards

Inventory management Process of buying, using and selling materials and products

Key performance indicator (KPI) An area within a business that can be measured in terms of a value, for example project completion times

Kinetic lifting Physical act of carrying, moving, lowering, pushing or lifting a load without the use of mechanical means

Lead time Period of time between ordering and receiving goods or materials

Legislation Current primary laws, sometimes known as Acts, created by UK legislative bodies (the UK Parliament, Scottish Parliament, Welsh Parliament and Northern Ireland Assembly)

Liability Legal responsibility

Linear measurement Distance between two given points along a line

Listed and heritage building regulations Formal guidelines that exist to protect and preserve buildings of special architectural or historical interest

Loadings Application of a mechanical load or force on a structure

Local Authority Building Control (LABC) Local authority department responsible for inspecting building work against building regulations and signing off completed projects

Lone working Employees working by themselves or without direct or close supervision

Malware Software designed to disrupt, damage or allow unauthorised access to a computer system

Manual handling Any lifting, carrying, supporting or moving of a load using bodily force

Metadata Set of data that describes or analyses other data

Metric units Decimal units of measurement based on the metre and the kilogram

Mobile bowsers Wheeled trailers fitted with a tank for carrying oil

Modular construction Combining factory-produced, pre-engineered units (or modules) to form major elements of a structure

National Grid Network of power lines supplying electrical energy around the UK

Networking Activity where businesses and people with a common interest meet to share information and develop contacts

Node points Intersections of lines or pathways in a diagram

Open questioning Asking questions that allow someone to give a free-form answer

Overheads Costs of running a business that are not directly related to production

Passive-aggressive Indirect behaviour expressing negative feelings and hostility

Permeable Porous, allowing water to drain through

Planning permission Approval that must be granted by the local authority for certain types of construction work

Prefabrication Construction of buildings and their components at a location other than the building site

Principal contractors Contractors appointed by a client to take the lead in planning, managing, monitoring and co-ordinating health and safety in a project involving more than one contractor

Procurement Process of agreeing business terms and acquiring goods, products or services from suppliers

Prohibition notice Legal document issued by the HSE to an employer that prevents work from continuing when there has been a serious breach of the law and people are at risk of immediate harm

Public limited company (PLC) Similar to a private limited company, the main difference being that money can be raised for the business through investors buying shares on the stock exchange

Ratio Relationship between two groups or amounts that expresses how much bigger one is than the other

Rebar Reinforced steel bar commonly used in concrete to act as a frame to stop it moving and cracking

Regulations Secondary laws made under the authority of the UK legislative bodies that created the primary laws; formal guidelines used to apply the principles of primary laws

Renewables Natural sources of energy, for example wind, tidal and solar

Renovation Alterations or improvements made to the fabric or structure of the interior or exterior of a building

Restoration Process of returning a building to its original condition

Risk assessment Process used to identify, control and record hazards in the workplace

Scale When accurate sizes of an object are reduced or enlarged by a stated amount

Screed Levelled layer of material (often sand and cement) applied to a floor or other surface

Section 106 A legal agreement between a local planning authority and a landowner/developer granting planning permission with obligations to support its aims of improving local services, e.g. creating and equipping a play area on a housing site

Self-employed State of working for oneself rather than an employer; a self-employed person is responsible for paying their own tax and National Insurance contributions on any earnings

Skin Single thickness masonry wall

Snagging Checking for minor faults in building work or materials

Software Sequence of digital instructions designed to operate a computer and perform specific tasks

Sound-absorbent Where sound waves are suppressed or absorbed by an item or a structure, rather than being reflected

Specifications Written documents that contain a detailed description of the materials, finishes, workmanship and construction of a building project

Statutory law Written law made by the UK Parliament; also known as an Act of Parliament

Stop blocks Physical barriers used to prevent construction vehicles falling into an excavation

Tenders Process of inviting bids from contractors to carry out specific projects

Testimonials Statements of recommendation produced by satisfied customers or clients that confirm the quality of a product or service

Thermosetting plastics Plastics that once formed cannot be reformed

Tolerances Allowable variations between specified measurements and actual measurements

Topography Physical features and shape of land surfaces

Topsoil Upper layer of soil, usually between 50 and 200mm deep, that contains most of the ground's nutrients and fertility

Transformers Devices used to change the voltage in one circuit to a different voltage in a second circuit

Tree preservation orders (TPOs) Legal protection of trees from damage, cutting, uprooting or removal

Triangulation Surveying method that measures the angles in a triangle formed by three survey control points

Trigonometry Branch of mathematics concerned with relationships between angles and ratios of lengths

Unexploded ordnance Explosive weapons that did not detonate when employed, for example Second World War bombs

Vernacular construction Construction methods sympathetic or particular to a region

Walkover survey Physical inspection of a building site

Water table Top of an underground level where groundwater permanently saturates the land

Water undertaker A water company that has the statutory duty to supply water/and or sewage services within a geographical area

Working capital Sum of money remaining after all the business's debts have been covered

Index

Check your understanding and progress at **www.hoddereducation.co.uk/myrevisionnotes**

My Revision Notes: Onsite Construction T Level